FAREWELL MA BELL

Enjoy & Best wishes

Donald Bross

Other books by Donald Bross

SURROGATE SON--a novel, is available either through your bookstore, the author or from the publisher, Book World, Inc.

MOUNTAIN MADNESS--a hard cover book-length true story to be released by New Horizon Press in September, 1996. Co-authored with Jimmy Dale Taylor. Look for it in your book store.

FAREWELL MA BELL

This is a true story. The events depicted in this book did happen basically as narrated, although certain scenes have been modified for clarification and to lead the reader down a smoother path.

Library of Congress Catalog Card Number:96-84971

ISBN 0-9652139-0-0

DEDICATION

This book is dedicated to the men and women who fought the wars.

To those Switchboard Operators who suffered through heat and job pressures to route calls.

To the Business Office Reps. who not only had to deal with a demanding public but had to suffer the stings and arrows thrown at them by the employees from other departments, specifically installation. Yes, I plead guilty.

To those who worked in Accounting and saw to it that we were paid on schedule.

To those who worked outside in what used to be the Plant Department; it was they who climbed the poles, crawled the manholes and often worked day and night to be certain service was established and maintained.

To those Engineers and Central Office personnel and all others who worked behind the scenes and without whose assistance we could never have provided service.

And to their spouses, who perhaps sacrificed the most. Thanks, Dixie.

CONTENTS

FOREWORD

Ma Bell was often misunderstood, loved, hated, used, abused, trusted and mistrusted. These emotions came from her million or so children. God alone knows what her customers thought of the old lady.

FAREWELL MA BELL is the story of one man's journey through the Bell System, and much more. It begins in the summer of 1951 and continues until her death on the last day of 1983. The reader will experience pole climbing, installing of telephones and twenty years in management. They will follow tornadoes, ice breaks, and hurricanes. Although much of the story takes place in Oklahoma, we see Ma in Texas and other surrounding areas, as well as New York State, Westchester County.

We see college grads change the System and the battles between these whiz kids and the old hands. We also see conflict between various departments, as each strives to enhance indexes that were designed to reflect work performances. Occasionally they did. We see the old girl plunge toward her own destruction. Although we blame the U. S. Government for her demise, we even wonder if she was just tired and wore out and might have even welcomed the end.

FAREWELL MA BELL is filled with anecdotes and laced with humor. Any person who has worked for the Bell System or who was a customer of the largest phone system in God's holy universe should want to turn a page and read on.

So, before we say farewell, let us begin with: HELLO MA BELL!

PROLOGUE

In many ways the Bell System was an anomaly with whom no other company upon our troubled earth could compare. Not only was she a mammoth monopoly, but Ma Bell was also the largest private employer in mankind's tumultuous history. Yeah, she was a breed apart.

Why the name, "Ma Bell"? Several answers come to mind. One usually feels secure with a mother who is a larger-than-life figure, with a parent who can provide financial, physical, and psychological stability. This was the image the old girl tried to project and most of us kids were content with her umbrella of security.

Also, a warm feeling comes from having sisters and brothers, even if there were some million or so to consider. Although we had our differences, even as do most blood relatives, there was quite often a closeness, a camaraderie that was difficult to explain to any person outside of the immediate Bell family.

There were times when we felt as though it was impossible for outsiders to understand us. Perhaps this is why so many of Ma's children often married another child, even though some had to first divorce an original mate to do so.

This is not to imply that we were always one big happy family. Far from it. What kindred doesn't have differences, petty feuds and sibling squabbles? And yet there was an invisible bond that seemed to bind us in some subtle way.

The policy of many corporations prohibits hiring more than one member from the same family. Ma Bell didn't adhere to that practice. She seemed to feel that if she had one good employee, others from the same clan would in all probability fall into the same niche. So it was not unusual for husband and wife, brother and sister, father and son, or mother and daughter to work in the same town and often in the same department. This hiring practice only added to the family feeling.

Now the old girl is dead. Her golden past lives only in the pages of history and memories of us who remember her days of glory. The eight companies who arose from the ashes of the once mighty Bell System are large corporations whose primary concern lies in profits and the bottom line.

Her children of long ago are losing their identity. Today there are folks with twenty or thirty years on the payroll who remember what it was like in Ma's glory days. Many of these employees no longer feel a special loyalty to their company, nor do they feel financially secure. The primary concern of many in this group is survival, to hang on until they reach retirement age, and then hope for the best.

From reliable sources we have heard of employees with twenty or more years with one of the so-called Baby Bell Companies whose jobs have been declared surplus. They have been told they have 90 days to find themselves another job within the Company or else they will be forced to take retirement, if eligible, or severance pay if not. The Company makes no attempt to help them find another slot; this is left entirely up to the employee. Ma Bell would never have treated her children in this manner.

No, she would not have.

Even some retirees are beginning to feel a sense of insecurity. When Ma Bell was alive, she went out of her way to care for these folks and to keep them informed. A Benefit Office was located within the State and there was always someone available to answer questions from those who were retired. Soon after divestiture, the Office for Southwestern Bell was centralized in St. Louis and now when you dial a number a machine answers your call. The word we hear is that these Benefit Office employees are going to be declared surplus and their job duties turned over to an independent contractor. At one time retirees were thought of as the men and women who had fought the wars and built the Company. Now there are some who seem to look upon these same folks as little more than a drag on the bottom line.

Not that retirees aren't being cared for; it's just that we are made to feel somewhat isolated.

As the world's largest corporation, Ma made a tantalizing target. For decades, it has been the nature of our Federal Government to make itself larger even while making corporations smaller, thus reducing any political threat. Many corporations can be kept in check by encouraging foreign competition. Since this plan of attack wasn't practical to use against the Bell System, Uncle Sam mobilized his Department of Justice and spared no expense. It's probable that he needed to do something to justify keeping those high paid attorneys on the payroll.

There is much ado these days about downsizing of Federal Government. It's difficult to believe that such a move will ever take place. More smoke and mirrors, more sleight of hand in an attempt to fool the electorate? We'll see.

Meanwhile, have fun, read the book and don't take life too seriously. Please indulge me if I write too much about me and not enough about our dear old Ma Bell. To the extent possible, I have tried to include other people. The chapter I enjoyed working the most is: OTHER VOICES, for in this many of you contributed. I believe that an entire book could be written in this manner. Perhaps that will be a future project.

A million stories could be written by those who have been a child of Ma's. This one was.

ONE

MA BELL'S CHILD

In 1876 several men were working toward the invention of the telephone and it seems as though more than one had reason to believe he was first to envision a talking machine that would actually work. On a morning in mid-February, Gardiner Hubbard, Bell's financier and future father-in-law, filed for a patent on Alexander's behalf, using drawings that featured undulatory current. Some four hours later, Elisha Gray's attorney arrived at the patent office and filed Mr. Gray's caveat for a speaking telephone. Had Mr. Gray's man come a little earlier in the day we might have worked for Southwestern Gray Telephone Company. Southwestern Gray? Naw, it just doesn't have the same ring to it, does it?

Before the summer of 1951, Mother Bell wasn't aware of my existence. I was nearly as ignorant about her.

During my formative years, the only telephones in our rural community were the antique, crank-type wall boxes, connected to the outside world by whoop and holler lines. One had to stand near a mouthpiece and shout, "Hello!" to have any chance of being heard. From our home the nearest instrument was some two miles by road, slightly less if one were willing to cross creeks and fields.

It wasn't until after World War II that dear old Ma decided to set poles and string wires near our home. True, her decision was reached after a less than gentle nudge from Uncle Sam who insisted that she serve all citizens, not just city dwellers. It seems as though Uncle always has enjoyed taking a swing at Ma. In fact, he kept jabbing and thrusting until the old girl was deader than a balanced Congressional budget.

More on that later. In 1951, at the venerable age of 75, she was not only alive but was getting her second wind.

Several years earlier, when I fit into the category of lanky farm boy, while trying to shock barley and watch the activity along the country road at the same time, I envied the young men who walked wood, pulled up wires from the ground and tied them to insulators. A few days later along came a telephone installer. It was a miracle. A two piece instrument, with crank, was installed on an eight-party line, meaning we could listen to the latest gossip and be quite sure others were listening to what we had to say.

At the impressionable age of sixteen, I had a sweetie, a town girl who eventually became my wife. A petite brunette, her name was Dixie Lee Ross and she was the prettiest little thing I'd ever seen. Being a town girl meant she lived some seven or eight miles from our farm home and so our time together was much too limited so far as I was concerned. The ready solution was for us to do a certain amount of telephone courting. And so we did. However, because seven other families shared our party line, privacy was a luxury we didn't enjoy.

One day, in lieu of being together, we were busy whispering sweet nothings into our mouthpieces when a third party lifted a receiver, listened, and hung up. This happened again, and then yet again.

This crusty old neighbor who had a love affair with four-letter words, the same gentleman who had said to me, "Your dad would have been better off had he thrown you into the creek the day you were born and bought a sow hog," listened long enough to hear more than he should have heard.

We were shocked into silence by a voice of doom that said, "If you want to see her so damn bad, why the hell don't you jump into your car and drive to town?"

Because I didn't have the time and probably didn't have money for gas, that's why. Did I explain this to the neighbor? No, heck, no! We were out of the mood to whisper more sweet nothings or anything else. "Good-bye." "Good-bye." Click! Click!

There are times when it seems as though fate controls ones destiny. When you miss a year of school through no fault of your own, you feel as though a mule has kicked you in the shins. Three weeks into the fourth grade I broke a leg and was out for the season.

When I returned to the country school where one beleaguered teacher taught grades one through eight, I was in a class with the *little* kids. This fact gnawed at me all the way through the eleventh grade. So, when my original classmates received their high school diplomas, I dropped out and took a wife.

She was the same, pretty little Dixie Lee who I had managed to woo and win. We rented an apartment and became expectant parents while I worked for Safeway in that faraway land, Oklahoma City, where I sacked groceries, stocked shelves, and often sorted, watered and made pretty the produce section.

Everything about that job suited me. Everything other than working indoors, being around multitudes of people every day and having to follow orders. There are times when farm boys are too independent. Stubborn, some would say.

Late that summer we had an opportunity to return to the country. Grandmother Bross, recently widowed, moved to town, leaving her house vacant. I'd been raised in a separate house on the same farm and my folks still lived there. To make ends meet, my dad was working at the Chevrolet Garage in Guthrie. Someone needed to work the ground and handle the livestock. When pappy asked me to do so, I jumped at the chance.

It's doubtful that my pretty little wife was overly enthused when we abandoned civilization so I could return to the soil. Still, she didn't complain when we moved from the luxury of a little apartment with bath to an old three room farm house where the fuel for heating and cooking was wood and our only bathroom was a number two galvanized tub along with a two-holer out back. Water was carried in from the windmill. Did we feel poor? Not too much. Even at this late date in our country's history, other folks were in similar straits.

Early on a cold winter night Dixie began to have labor pains. If ever there was a time to panic, this was it. We were seven miles from the hospital, and there was no way we would ever make it in time.

"Are you all right?" I asked repeatedly, as I all but carried her to our old Ford.

"Yes, I'm all right. We have plenty of time."

Little did she know. As I gunned the engine I had visions of a car-born baby. With only me there to help. To my surprise there were still only the two of us visible in the car when I parked out front of the hospital. We hurried inside.

When I reported to the receptionist that my wife was about to have her baby, I expected them to abandon the other patients and run to her rescue. Call the doctor and get him here in a hurry. The hospital folks remained so calm that I wondered if they'd ever done this before.

Eventually they escorted Dixie into an examining room and ordered me to stay out. Down the corridor was a waiting area. I settled into a chair and tortured nervous fingers by twisting them as I watched and listened to a wall clock jump ahead, one noisy minute at a time. Then I counted perforations in the ceiling tiles. What in the heck was going on?

Hours later the doctor arrived. He didn't look as though he was wide enough awake to deliver a baby. He examined Dixie, went into another room, and fell asleep. Dereliction of duty, I decided.

By the time they awoke the doctor, the night was nearly spent. I had left the waiting room and was pacing the hall. "Come with me and hold your wife's hand," said Doctor LeHew.

Who, me? I gulped and followed him in.

Baby Ted was born, adding more responsibility to a couple of kids new to the family business.

For one year we somehow survived while I carried on the family tradition of farming a quarter section of land, milking cows, separating cream, raising chickens and hogs, baling hay, planting corn, wheat and barley, and cutting firewood on the creek. A Farmall tractor, so pitifully small when compared to the enclosed cab, air-conditioned, stereo equipped, mammoth machines that now till the soil, was my salvation.

For me the environment was tolerable since it was much the same as what I'd experienced since childhood. For Dixie it must have been comparable to stepping back into the nineteenth century. Winter or summer, she cooked on a wood stove. No instant hot water, no air conditioning, not even a ceiling fan. Stuck out in the country and her not knowing how to drive. However, the pretty little town girl gritted her teeth and came through like a real trooper.

My farm career was over almost before it started. In the summer of 1951, Grandmother Bross decided to sell her land. Wild thoughts of buying it ran through my mind but having few nickels to call my own, this was not possible. The time of the small tenant farmer being able to survive in these United States was going the way of the covered wagon. So, responsible now for a wife and baby and having only an eleventh grade education, what I needed, and needed in a hurry, was a job.

Safeway was a fine organization but I had no desire to work a job that kept me indoors. Besides, I was not at all sure they would hire me back. Once you quit, most companies won't. While I was mulling over the situation, my dad offered a possible solution.

Shortly before The Great Depression hit, he had graduated from high school and hoped to work for *the* Telephone Company. Unfortunately they weren't hiring, so he took the only avenue open at the time and became a tenant farmer. In late July of 1951, we were under a spreading elm, eating a homegrown watermelon, when he said, "Why don't you try to get on with the Telephone Company?"

"They wouldn't hire me."

"You won't know until you try. Son, you need to get on with a company where you'll have some benefits. You don't want to be stuck with work like I'm doing for the rest of your life."

I took his advice.

The next step was one I dreaded. If I was to have any hope of working for Southwestern Bell, I would have to enter into the big city and put in an application.

Although our farm lay only twenty-five miles from the heart of Oklahoma City, to me it was like entering into another world when I drove into that sprawling metropolis. Downtown was a jungle. People, people, everywhere, and not a familiar face to be seen. Drivers gone wild. Buses belched smelly fumes.

Other than for the few miserable months we had lived within its limits while working for Safeway, my trips to the big city had been rare. Oh, there'd been a few visits to the State Fair where I'd been more interested in the rides and the sideshows than in livestock exhibits. Heck, I could see plenty of those critters at home.

Occasionally we had hauled calves or hogs to the stockyards in a little two-wheel trailer. Packing Town, as we'd called it.

When Dixie and I were courting, we'd dared a time or two to venture into this unknown land. One Sunday afternoon I drove us down to see my singing idol, Mr. Hank Williams (the old man, not the kid) in person, and he did manage to stagger onto the stage at the Municipal Auditorium, almost sober, and, in his own unique way, give a magnificent performance.

Also, there was the time when we skipped school and, for some reason that is no longer clear, headed for downtown Oklahoma City. I parked the car where it was not supposed to be parked, and then we walked around. Upon returning to the car, we found a parking ticket tucked under a wiper blade. When you're seventeen and in a strange land, this causes a sinking feeling in your belly.

My first thought was to ignore the thing, but then I had visions of the city police contacting my folks and alerting them to the fact that their eldest son was playing hooky from school instead of hitting the books. So, we headed over to the police station.

We were standing in the ticket-paying line when two bruisers in plain-clothes, big enough to play tackle for the Chicago Bears, made me the meat of a human sandwich. Without explanation, they escorted me into a small room and sat me down, leaving Dixie to wonder what in the heck was going on. Yeah, I kind of wondered too.

Was I scared? You bet I was. I hadn't realized they played so rough just because you skipped school and got a parking ticket.

Eventually their reason for the pinch came to light. They had a grainy photo of some outlaw from Massachusetts who'd murdered a few folks and was thought to have headed west in hopes of eluding his pursuers. These detectives thought my mug matched his. Don't know why they ever thought that. Personally I thought he was an ugly hombre.

Anyway, there we were, the desperado who was sweating it out while his pretty little moll stood by in another room. Before they started with the rubber hoses,

I managed to convince them that I was just a poor, dumb little old farm boy who had never been east of Arkansas. They were sorely disappointed that I was not a mass murderer. Could be that I cost them a promotion.

Oh, well. That was the last time we ever skipped school.

Having decided to take my dad's advice, I climbed into my vintage, vapor-locking '36 Ford, drove to the big city and ended up in Ma Bell's Personnel Office. I filled out an application, then took a test. Compared to those required today, it was simple enough. After completion of these chores, I was escorted into an office where I was introduced to Mr. Fuller.

He was a pleasant fellow and evidently needing help in the worst way because in spite of roadblocks that I inadvertently erected, he would eventually welcome me into the fold. There was a time when I believe he would have preferred throwing me out the window.

Strike one against yours truly was having only an eleventh grade education. "If I hire you, will you finish high school?" asked Mr. Fuller.

With no idea of how I'd ever manage to do so, I promised that I would. Miracles do happen. Seven years later I had a diploma and after that managed two years of college at Oklahoma U.

Strike two was called during the physical exam. Operations on the broken leg those many years ago had left a rather nasty looking scar. The doctor took one look, frowned, and shook his head. "You wouldn't last long climbing poles," he said.

"It doesn't bother me a bit," I insisted.

It's doubtful that he was fully convinced but he did eventually give his stamp of approval.

With the results of the physical in hand, Mr. Fuller had to make a decision. By this time all that was in my favor was that he preferred hiring farm boys for certain jobs that required muscles developed through years of physical labor. He was looking for someone to put in a day's work, not to invent transistors or laser optics. Bell Lab had some educated folks on the payroll who would patent these revolutionary marvels in days to come. And so what he offered me was a job as a located lineman in Oklahoma City.

My reply was, "Mr. Fuller, I'd like to work for you, but I can't stand the thought of having to live in a big city again."

Any way you look at it, that should have been, "Strike three; you're out!"

Had I been him, I would have told the brash farm boy with an eleventh grade education and a potentially bum leg to leave now and never darken my door again. However, Mr. Fuller was a nice man.

After a sigh and in all likelihood much more than a little soul-searching, he offered me a job as an unlocated lineman or, if that wasn't suitable, I could be located in Clinton. If I chose to be unlocated, I would be expected to work anywhere in Oklahoma's Western Division, a territory that covered some three-fourths of the state. It would at times be necessary to move on about five minutes notice.

Being a family man, I opted for a located title. He failed to mention that located linemen outside the metro areas spent many a night in an old hotel. An oversight, I'm sure.

So there I was, about to become a child of Ma Bell's. Mr. Fuller wanted me to report to work in Clinton on Monday, the sixth of August. For reasons of great importance, reasons that I can no longer recall, I advised him that I couldn't possibly start until Wednesday, the eighth day of August.

Strike four!

Was that gnashing of teeth I heard?

How could he tolerate such ingratitude?

Mr. Fuller was a nice man.

In 1951 Southwestern Bell had archaic rules. Yes, they did. You were not permitted to drive your personal car on company time. Period. If you had an accident, the company could be liable and even find themselves in court facing a belligerent victim. I was not allowed to drive my old Ford to Clinton. Since I would be on the payroll, earning ninety three and three-quarter cents per hour, I had to take the bus, the cost to be reimbursed.

Around eleven in the morning, as the thermometer made its inexorable climb towards the one hundred mark, the bus pulled into Clinton. I had the address for Glenn Wilkerson, the District Construction Foreman over the Clinton District. Following several inquiries I found his office, two small rooms located on the second floor of an ancient downtown building.

Soon after introducing myself to his clerk, I was admitted into the Great Man's lair. It was a little room with wood-sash windows open to the street and an oscillating fan moving the hot air (coming not from Mr. Wilkerson but from the summer sun. For the most part, anyway).

Even as we eyed one another and shook hands, I felt intimidated. Here was a tall man, middle-aged, who seemed old when compared to my youth, a man who was impeccably dressed. Later I heard him called, behind his back of course, "Slick" or "Diamond Jim". He did wear a big ring with a large, glittering rock. It could have been a diamond, I suppose.

He spent a half hour or so praising the Telephone Company. If I kept my nose clean and worked my fingers to the bone, I'd never get rich but would make a decent living. I was impressed by his intelligence and somewhat awed by his grandiose demeanor. Years later, when Slick retired, I heard that he became Mayor of Clinton.

He then identified my immediate supervisor and where I was to report. Byron "Boog" Latimer would be my boss. He and his crew were working at Sayre. I was to catch the next bus west.

Hold 'er now! What's going on? What happens if your wife doesn't drive? Dixie didn't.

"Mr. Wilkerson," I said, "I was hired as a located lineman, to work in Clinton."

He shot me an impatient look. "Just because you have a located title doesn't mean there won't be times when you'll be working out of town. When required to do so, the Company will pay for your room and board."

"Mr. Fuller said I would be working in Clinton."

"No, no. You must have misunderstood."

"No, I didn't misunderstand."

"There will be times when you'll have to work out of town. This is one of them."

We argued about this until near noon. As I later recalled those moments, I realized that he was close to telling me that he and his Company had no use for such stubborn farm folks, and that he was quite close to buying me a bus ticket home. I told him I wanted to call and discuss this new development with my wife.

Perhaps I'd underestimated his patience since he agreed to this. However as I later discovered, he was having trouble keeping folks on his payroll. He neither invited me to lunch nor offered me the luxury of using his company telephone. Simply told me to report back at one o'clock with my decision.

I found a drug store with a lunch counter and a pay phone. I called home and told the sad news to Dixie. Our first inclination was to forget Ma Bell. But as we continued to lament our pitiful financial state, we decided that after all we did need a job. So, we would give it a try.

I was near starved. When odors of ham and cheese sandwiches tickled my taste buds I dipped into my funds and ate one, along with a malt. Then I returned to Mr. Wilkerson's office and gave him my decision.

For better or worse, I was now one of Ma Bell's million or so children. Soon I was on a bus, heading west to Sayre. Five strikes, and I still wasn't out. Yeah, they must have needed help in a bad way.

TWO

WALKING WOOD

It was a time when certain jobs were performed by men and others only by women. At Ma Bell, men climbed the poles and performed the technical work while women worked as operators and handled the Business Office. Tradition argued against change.

Open-wire lines ran from coast to coast and even beyond border to border. Communication satellites were far in the future. Many of us felt as though innovation had advanced to its limits. Little did we know.

Mother Bell wasn't the only company to benefit from technology. Today it would be most difficult, even impossible, to find a cross-country bus in the United States that doesn't provide air-conditioning.

With side windows open, we rolled west on Highway 66, over the same road that Steinbeck had taken his Joad family as they moved towards their Promised Land. Upon arrival, they discovered only Grapes of Wrath.

Nearby, Custer and his soldiers had fought the Battle of the Washita, where they swooped down upon a sleeping camp, killing scores of Indians. Heck, they even managed to wipe out a Brave or two.

Custer, of course, was soon to be paid in full. On that same June Sunday in 1876, when General (Colonel at the time) George Armstrong was dispatched to soldier eternity by other pesky Redskins, far away, in a slightly less hostile environment, A. G. Bell was at the Centennial Exposition in Philadelphia. Mr. Bell tried continually to convince folks that they should take his far-speaking box seriously. This was a most difficult sell. Some thought he must be promoting the work of the devil. Fortunately, the judges came in alone on Sunday morning and were quite impressed. This made Mr. Bell's day.

After leaving the bus at Sayre and making a false turn or two I found the reporting place for the construction crews that were then working out of Sayre. It wasn't a work center at all, but a space leased along the railroad tracks large enough to accommodate wire and poles and crossarms and a shed where brackets, insulators, and other paraphernalia needed for country construction of open wire lines was stored.

Neither Boog nor the welcome wagon awaited me,. There was only a young man, a stockman, who was busy unpacking material.

"Where's Mr. Latimer?" I asked. Without looking up, he said, "He should be here about four-thirty."

Sure enough, a few minutes before quitting time, in roared four trucks, five-man cabs each, and rolled to a stop. It was customary for the foreman to ride up front with the driver and for linemen other than the driver to occupy the back seat.

Out of these trucks poured men (only men in those days), all young, other than the four foremen who seen through my young eyes looked ancient. Shoot, some of 'em must have been past forty.

To the man they wore boots, hats or caps, long-sleeved shirts, safety glasses, and gloves. These items were all mandatory garb for Bell System linemen, winter or summer. Any nonconformist would soon see his name struck from the payroll.

They seemed good-natured. Even though they had put in a hard day's work, being young, many of them single, some were looking forward to a hot time in the old town tonight.

Other than for Mr. Latimer, the foremen were a serious lot. Old Boog, tall, rawboned, rust colored hair, and red faced, had a mischievous aura about him. He looked me over, shook my hand and said, "So you're my new lineman, are you?"

Lineman? Me? "Yes sir, I am."

"Let's go get you checked in and I'll tell you what the rules are."

He escorted me to an old hotel that must have dated back to shortly before Custer, checked me in, and later directed me to the cafe where I would eat twice a day. Breakfast and supper, that is. For the noon meal, they fixed us a sack lunch. I was still surprised that the company paid my expenses while I was away from Clinton. This seemed like a pretty good deal after all.

As for the food, we couldn't order steak and/or lobster. It's doubtful there had ever been a lobster inside that place anyway. This was in the days before the fast-food chains when home-cooked meals were the norm instead of the exception. We all ate at the same cafe where we signed our meal tickets and it was left up to Boog to settle the account. Even so, we weren't given carte blanche to order whatsoever our heart or hungry belly might desire.

"Your meals are not to average more than eighty-five cents each," Boog said.

Woe to the man who exceeded the limit, for whatever reason. He would then see the other side of Boog, the one who could scald you with invectives. Off the job and away from Company business, that rascal was an amicable sort of fellow. At other times he could be a totalitarian. Whatever folks might have said about him over the years, and of course some of these remarks were not all that complimentary, the company got their money's worth out of old Boog.

The following morning I was at the work yard and it was soon eight o'clock. At the magic hour it was as though someone had flipped a switch. The men, some who had been jawing with one another, some who were exaggerating their conquest of a native girl, some with hangovers, swung into action. Now it was time for serious business. Time to earn your paycheck. Eight hours work for eight hours pay was a slogan I was to hear often in the years to come. In time, I was saying it with conviction to a few folks myself.

It's kind of hard to argue with the logic, isn't it? Anyway, it should be.

We loaded the trailer with wire and the truck with supplies. On the way we stopped at the icehouse to fill the water jug, and then headed for the country. Several miles out we unloaded and went to work.

The Bell System was nearing completion on its goal of a phone in every home, so ordered by Uncle Sam. Revenues from country folks would never pay for poles, wire, and labor. The deficit would have to be paid by the big-city customers, pay-phone users and long-distance revenues. Therefore, Ma had to be pushed a little.

Even as they'd come to our country home several years earlier, now we were providing more than whoop and holler service to those good country folks outside the town of Sayre.

Of course I didn't do too much of the providing.

I spent that first day looking busy, attempting to grasp the work operations, learning the terminology, pulling wires off of reels on the trailer, and reacting when Boog barked. He barked often.

Gone was the jovial man I'd met the day before. The happy fellow made an appearance during our thirty minute lunch break and didn't show his face again until the four-thirty quitting time.

Friday, my third day on the payroll and second on the job, Boog decided if ever I was going to earn my pay it was time for me to strap on the harness and move up in the world. In the years to come, before a man (or woman) could climb their first pole, they had to first pass a training course with safety being the foremost consideration. Even then, the Bell System Safety Slogan, strictly adhered to for the most part, was:

No job is so important
and no service is so urgent
that we cannot take time
to perform our work safely.

Boog treated that slogan as though it had been brought down from the mountain by Moses. When you opened the small door that concealed the water jug, it was there for all to see.

Even though Boog must have realized a potential hazard awaited, he had to make do with what he had on hand. He pulled belts and climbers out of a truck bin and kept experimenting until he found some that came close to fitting. Satisfied, he led me to the nearest pole and began his class.

"You don't ever walk on the ground wearing your climbers," he said. "Might cut yourself with a gaff or dull a point on a rock."

One step from the pole was the maximum allowed.

According to stories I've heard, this concern for safety was relatively new. Back in the olden days, when men were real men and wood-walkers were elite, they strapped them on in the morning and didn't take them off until night. If then. Could be that some folks wore them while asleep.

First, under Boog's scrutinizing eye and listening to emphatic instructions, I stood on the ground, placed the safety strap around the pole, and then leaned back. Nothing to it. Next came the function of jabbing a gaff into the pole and lifting the other foot until only a tiny, tapered point of sharp steel penetrated wood and supported my weight. When both were firmly planted, I unhooked one snap of my safety strap, passed it around the pole from left hand to right, and snapped it into the large "D" ring. Feel. Listen. Look. Only when you have done all three do you relax.

When climbing, maintain three-point contact at all times. So said Boog. Emphatically. Either two hands and one gaff on the pole or two gaffs and one hand. When safetied off, two gaffs and a safety strap would suffice.

The following week, upon having reached working height for the first time, I opted for four-point contact: two gaffs, one safety strap, and, for good measure, one hand. This seemed to me to be a sensible approach. So far as safety was concerned, it beat the other method, hands down.

Boog would have none of it. He shouted, "The Company's paying for both of your hands. Use 'em!"

It's difficult to turn loose that first time.

Back to my initiation. Take two steps up. Now three. Climbing ever higher while Boog is on the ground, bellowing instructions. "Don't look at your feet! Look up!" Ten feet in the air. "Safety off and lean back."

Yeah, you bet. With fumbling fingers you pass one end of the safety strap around the pole and feel the big ring, listen for the snap, and look just to be sure. Only then do you lean back. After giving yourself a good talking to and convincing yourself that you are in no immediate danger of becoming a statistic, you turn loose.

Enough for one day. Quitting time. Gather at the truck. Secure the wire reels. Head back to Sayre. Catch a bus for home. Since I was traveling on my own time, I had to buy the ticket. I was experiencing a heady feeling. For sure, I would have something to tell.

By golly, I was a wood-walker.

The next Sunday I drove to Sayre and checked into the hotel. I was expected to be ready for work at eight Monday morning, wide-eyed and alert enough so that I wouldn't be a danger to myself or to others.

Come morning, my fourth day on the job, I was anxious to polish my wood-walking skills. A good deal of polishing was still needed.

I was some surprised to discover that I was no longer junior man on the crew. Five days had passed since I had ridden the bus to Sayre and already I was off of the bottom so far as seniority was concerned.

Our newest member was Lloyd Blasingame. At twenty-seven, he was one year past the upper age limit for new employees in those days.

Before the government decreed it unlawful to discriminate in hiring because of age, sex, faith, color, race, freckles, or general dislikes, it was Company policy that new-hires be between the ages of eighteen and twenty-six. From their viewpoint, no one should work for the Bell System unless he or she intended to stick around until retirement age or until death do us part. Of course Ma was thinking of the employee's death, not her own.

If they were going to eventually pay a pension, they wanted at least forty years of hard labor before the first check was mailed.

Alas, many of those hired as linemen deserted Ma Bell some thirty-nine years short of pension eligibility. Some found they preferred keeping both feet firmly on the ground; others wanted to plant a foot firmly up Boog's you-know-what. Not only Boog's but others of his ilk as well.

Since there was a constant turnover of personnel, it was becoming increasingly difficult to fill the positions when vacancies occurred. Therefore, the rules were bent slightly. Lloyd, an old man of twenty-seven was hired. He was cheerful and witty, the type of guy you couldn't help but like. From our first day together, we became friends.

Later that week, I faced a tough decision. The word *Union*, as in *Labor Union*, was a word seldom spoken back home. Having been raised on a farm, the only man I'd had to answer to, other than teachers at school and my boss at Safeway, was my dad. It had never occurred to me to set up a picket in protest of low pay, long hours, or poor working conditions.

What little I had heard about Unions was about John L. Lewis and his striking coal miners. In our family he was held in about the same esteem as Hitler or maybe Mussolini. So when the President of Local 6014 of The Communication Workers of America traveled from Oklahoma City to Sayre to sign up those newly hired and to hear a grievance or two, I found myself in a dilemma.

We'd arrived at the work yard at the end of a hard, hot day and there was B. A. Floyd. Mr. Floyd was hailed by the old hands. It was soon apparent that he was well liked and respected. Back in the big city, B. A. was a cable splicer by trade. When he needed to conduct business for the Union, he traveled at their expense.

Come four-thirty, I was ready to make a break for the hotel but B. A. headed me off. A soft spoken man, he wasn't pushy or overly vociferous as was our impression of John L. He was a darn likable guy, and in the years ahead I came to value his friendship.

He made his sales pitch. The benefits were many and the dues were only three dollars a month. This now seems a pittance but for a man earning thirty-seven dollars and fifty cents a week, before deducts, and also supporting a wife and child, this seemed a considerable sum. When I vacillated, B. A. was persuasive.

Other members took me aside and offered their counsel. They didn't openly threaten, but I had a feeling that my tires might be slashed or sugar could be poured into my gas tank if I refused to sign. For sure I would be about as popular among my coworkers as the Pope at a Southern Baptist Convention if the card didn't bear my signature.

"We have to stick together," one man said. "It's the only way to get along.".

To keep peace and harmony, I signed the card. Although never a militant, a few years later I became an active member and was for a while Vice-President of the Local.

It was time to move my family to Sayre so we rented a one-room furnished apartment. The forty hour work week gave us time together. Ted, our little boy, learned to walk on the tile-covered, concrete floor. Of course he received a bump or two to the head. Not much damage was done to the floor.

Soon the great rural program was nearing an end. We packed up and moved to Clinton.

THREE

LOCATED AT LAST

An upstairs apartment in an old two-story frame house became our home when we moved from Sayre to Clinton. The bathroom, shared with three other families, was at one end of the hall. Inconvenient, yes, but a step or two ahead of a galvanized tub and a two-holer.

All in all it was about as much comfort as we could hope for. Dixie needed a little comfort. She was with child again.

Clinton was a pleasant town, surrounded by family farms and ranches. It was a little larger than Guthrie and seemed to be filled with friendly folks.

I had spent my childhood in what had once been the Unclaimed Lands of Indian Territory. Located near the center of the state, it seemed as though none of the Civilized (Nor had the uncivilized, come to think of it) Tribes really wanted the five counties that were later born from the Run of 1889. The whites and blacks filled up nearly every nook and cranny. Seldom did we ever encounter any pesky Redskins.

Not so in Clinton. Here they roamed the streets with a wistful look in their eyes, as though wondering how they could have let it all slip away. Some would be downtown, wrapped in a blanket of many colors that was reminiscent of Joseph's coat.

Or Dolly Parton's.

Although Dixie is part Indian, she was raised by her white family and so she never took up the custom of scalping enemies or sewing buffalo hides.

After moving to Clinton, we met Boog's wife. She was even more pleasant all the time than he was off the job. You couldn't help but like Daisy. To her we were just a couple of kids a long way from home, and she tried to make us feel welcome.

Boog was happy to be home. He seemed to mellow some, even on company time. After hours, he could be more fun than a clown at a Shriner's circus. Occasionally, several of us would gather at a downtown pool hall and play a game or two. If invited, Boog would join in. When thrown together in a small town, you fail to realize that management folks are a breed apart and that you really should keep your distance.

When the Red Cross came to town hoping to draw blood from all the able-bodied, ambulatory citizens, we on Boog's crew decided to do our patriotic duty and so we all went in to donate. When I tried, they wouldn't consider drawing my blood because I wasn't yet of age. I was still two years shy of being twenty-one.

I was married, one kid at home, another on the way, and they wouldn't take my blood. When I made the foolish mistake of protesting, this adamant lady insisted it would be necessary for a parent or guardian to sign a consent form before they could insert a needle.

Boog stepped up boldly and said to the lady, "I'm Don's guardian." Not having to provide proof, he signed on the appropriate line and that settled that. When they punctured my vein, I questioned the wisdom of one who is overly persistent.

There were times when the most unexpected event happened. One autumn day we headed for the rural area, some four or five miles out of town, where we would transfer two crossarms of open wire from an old pole to a new one. Lloyd was driving the truck. Two other men on the crew were tops in seniority. Close friends, they were skillful linemen but a little hot-headed. It wasn't unusual for the two of them to clash with Boog.

Our work site was about half way between two crossroads. In accordance with our safety practices, a Men-Working sign with a red warning flag flying from the top, would be placed at either section line. When we reached the first intersection before the pole, we stopped and placed a sign. Then Lloyd drove on to the job.

If there was one thing old Boog could not tolerate, it was for a man to stand around with nothing to do. Only the boss was to ever enjoy that luxury. You'd better be going through the motions, even if it was make-work. Especially if Slick (Mr. Wilkerson) happened to be out and about in his constant search for goof-offs. The thought of such a visit was enough to send Boog into convulsions.

Boog had Lloyd stop at the pole, then told the two buddies to harness up and walk wood. Lloyd was to assist them from the ground while I carried sign and flag some half mile farther along and placed them at the next crossroads.

As I was hurrying back, it became obvious that all was not well. The buddies unhooked their safety straps and all but fell to the ground. They shucked their tools, grabbed lunch buckets off the truck and began hoofing it towards town. When I arrived at the truck, Boog barked, "Get your tools on. You and Lloyd finish the job."

As we worked the pole and kept one eye on Boog, who spent the day mumbling to himself, I asked, "What in the heck happened?"

"They didn't like for Boog to stand on the ground and tell them every move to make," Lloyd whispered. "It turned into a shouting match."

They quit. In the twinkling of an eye, Ma lost two more children. Other than for having tempers they sometimes could not control, they were good men. Having three months service, I was now high in seniority.

As days passed, it became clear why Lloyd, at the age of twenty-seven, and me, with a limited education and potential physical problems, had been hired. Ma's new kids kept divorcing her and moving on. Although Ma offered security, like a dominating mother, she often wanted to control your life.

The seasons were changing. During the first cold snap each autumn, we used to say, "Winter is coming, and I done spent my summer's wages." In the fall and winter of 1951, we didn't have many wages to spend. We had no need for a checking account since we lived out of pocket, from one payday to the next.

One Sunday, only five days before our biweekly check, I had in my pocket only twenty-two dollars and change, this being the extent of our fortune and savings. Feeling affluent, I decided to take the family to Sunday dinner. We drove to a cafe where they specialized in home cooked meals.

We were fed full for less than two dollars. I left the waitress a dime or two and paid the bill. Within my lean wallet was our fortune, two ones and a twenty. I handed two bills to the cashier, received a few pennies change, and we headed for home.

Later that afternoon, for some reason, I decided to count my money. When that's all you have, you have a tendency to peek now and then to see that all is well.

All was not well. A dollar bill remained, not the twenty that was supposed to be tucked away. It seems as though I had made the mistake of handing over our only twenty in place of a one.

It can be devastating to find yourself down to your last dollar, especially when you have a family. Still, I wasn't overly concerned. Not then. We lived in a town filled with friendly folks and there was no doubt but what the people at the cafe were in possession of my nineteen dollars.

Ask and ye shall receive.

Within five minutes I was at the cafe, telling my sad story to the manager. He heard me out, then said, "There's no way I have of knowing for sure that's what happened until I count today's receipts. Come back tomorrow. If I check out nineteen dollars long, it's yours."

He lied.

Next day after work I hurried to the cafe. With a straight face, the manager said, "Sorry, but I don't have your money. We checked out to the penny."

After a lengthy argument, he ordered me out. I knew he was lying and he knew he was lying. Sometimes we're helpless to prove what we know

We were down to our last dollar until payday. Most of it went for the baby's milk, along with a loaf or two of bread. Fortunately we had a few cans of Spam squirreled away.

The work varied. In those days most of our cross country toll-lines consisted of bare, copper wires tied to insulators on crossarms attached to poles. Often these ran along railroad right-of-way as did Line-3, laying east and west, and maybe running from coast to coast for all I know. My knowledge was limited to a stretch between the towns of Clinton and Hydro. Much of our time was spent doing maintenance on this line.

The wire followed a railroad track and more often than not we couldn't drive in. Where we parked was usually a good distance from where we needed to work. Periodically Telephone Engineers walked the line and made note of poles, crossarms or insulators that needed replacing. It was common for us to load material onto a trailer, pull as close as possible to where we needed to work, and then walk and carry. Linemen took their turn working the ground and climbing poles.

My first experience with this was a little east of Weatherford. We parked truck and trailer and unloaded crossarms along the road. Since it was my day to do the donkey work Boog gave me a list of poles where these heavy, ten-foot long chunks of creosoted lumber were to be replaced.

While the fortunate linemen made ready to work aloft, I picked up a crossarm, weighing between forty and sixty pounds, wet creosote dripping like a runny nose, threw it over my left shoulder, and headed down the railroad track.

Boog called me back. "What's wrong with your other arm?" he asked.

As I stood there with what must have been a quizzical look on my face, he placed a crossarm on my other shoulder. Somewhat sadistically, he said, "That'll keep you balanced. Besides, the Company's paying for both your hands, not just one."

Some sense of humor.

Mr. Wilkerson believed in spending as much time as possible in the field. Once, when he came out to observe, I was up a toll-line pole and somewhat entangled in my work.

In those days I had to step heavy on the scales to weigh a hundred and forty pounds (This has changed considerably). Since I measured about six feet, I was between lean and skinny. It was a constant struggle to keep my body belt from sliding over my hip bones and right on down. Then I would have to take time out and pull it back into its proper place.

Anyway, I was up this pole, being observed from below by both Slick and Boog. They were two men who did a lot of observing. While standing on the lowest crossarm with my form laced through that mess of wires, I stretched out for the end pin, old number ten itself. That's when I experienced this sickening feeling as I felt my body belt slide over my hips, past my knees and right on down to my ankles. Nothing to do then but step out, unsnap and drop the entire apparatus to the ground; then it was climb down after it and harness up again.

Slick had seen enough. He shook his head and headed for his car. Boog called me over.

"You sure know how to make me feel good when my boss is around," he said.

"Sorry about that."

"Come over here."

He stuffed rags into my pockets. "Maybe that'll keep your belt up," he said with disgust.

It helped.

There was very little difference between being an unlocated lineman or having a located title in a town the size of Clinton. The weekly wage was practically the same. Only thing is: when you were working at home the Company didn't have to pay your room and board.

There was a Company rule that, although it must seem archaic in modern times, was strictly enforced in those days when Ma tried to control her children, on and off the job. If the town where you would be working for two days, a week, or whatever, was more than a thirty minute drive from your home base, the truck would be parked in that town and the company would put you up in a hotel and feed you for the duration. If you were still in that town come Friday, you had the option of staying or going home for the weekend, so long as you returned to the hotel by Sunday evening.

Other days, you could not drive home and spend the night. Not even on your own time. Their reasoning was: they wanted you well rested and ready for a full day's work, not fatigued from the rigors of driving to and fro. It was also forbidden to live more than a thirty minute drive from your located town.

Years later that edict was abolished or ignored. Some folks would drive an hour or two each way and nothing was ever said.

There were times when we would load suitcases onto the truck and be gone for a night or perhaps a week. One Monday we left home and stopped at small towns to do a little job or two and then moved on. Our destination was the Oklahoma Panhandle. By the time we arrived, we seemed to be half way across the country. Sparsely populated, this is good land but in places seems to be about as desolate as the moon.

It seems as though every tree in town was alive and well and living under our toll-line. We trimmed these and did maintenance work on rural lines. On Thursday we were in the country replacing brackets and insulators on a two-wire line and, as quitting time approached, still had a long list to work. Boog informed us that if we didn't finish that day we would have to do so Friday; then it would be too late to drive home. We'd have to stay until the following Monday.

None of us liked the sound of that.

It was spitting snow. We had visions of a real, old fashioned Panhandle blizzard where we would be drifted in until spring.

"It's up to y'all," he said. "If you want to finish today, we'll head for home in the morning."

The drive home would take all day. If we chose to finish, he couldn't pay overtime. Boog made it sound as though it was of no concern to him: he was making this generous offer only as a convenience for us.

It didn't make sense that the Company could pay our room and board for the weekend and yet not pay us for working a few hours overtime. Such was Company policy. Or perhaps it was only Boog's policy.

We opted to finish. We drove to the hotel after dark and headed for home the next morning. Boog could get the Company's money's worth out of Ma's children, that's for sure. That was my last long trip away from home while working at Clinton.

Safety was a primary concern. You were required to follow Company rules even if it killed you. Any man prone to injuries best be checking up on welfare benefits or placing his application at some mom and pop grocery store. Ma Bell would not tolerate such.

About all that would happen to a foreman if a crew member had an accident was that he would be in the doghouse with his superiors until such time as they forgave or forgot. When it came to accidents, those in the upper echelon weren't the forgiving sort and they had memories that would put an elephant to shame.

The sad foreman who allowed an on the job injury could forget all about promotions, pay raises or any of the goodies that might come the way of an accident free supervisor. From the Company's viewpoint, all accidents, other than when God intervened, were avoidable.

Debatable to us linemen was what to do if our climbers cut out while climbing. We had two ready options: either kick free and hit the ground, or, try to bear-hug the pole. The consensus among us, based on no hard evidence of course, was that the alternative least likely to cause serious injury was to kick free and hope for the best. That is unless working over concrete slabs, picket fences or other items upon which you might be splattered or impaled, and then you should by all means hug the pole. Quite early in my woodwalking career I decided that if this should happen I would kick free and land on my feet.

Come spring we were outside of Clinton, doing maintenance work on a telephone line within the grounds of the TB Sanitarium. A pair of wires hung on skinny little poles, scattered about the yard. One pole I needed to walk was about as big around as a broom handle and had been stuck so many times with gaffs that it would have been difficult to find a solid piece as large as a postage stamp.

No job is so important
and no service is so urgent....

What the heck? Other folks had climbed to the top. Before me was the evidence: a chewed up pole. Confident that I could climb the darn thing and change out a broken insulator, I started up. Working height was some twenty feet above the ground. It was like jabbing at a fence post. And then, about ten or twelve feet up, it happened: my climber gaff cut out.

Did I kick free and land on my feet? No! Did I kick free and land on anything? No! At a time like that, you don't think, you act. When you know you're about to free-fall, you react and there is no possible way that you can not do so. There is no time for conscious thought.

I bear-hugged that little stick and slid down about five feet. By that time my shirt was ripped to shreds and my belly, chest and arms were filled with splinters.

Boog hurried over, loaded me into the truck and rushed me to a doctor or to the emergency room. Right?

Wrong! Boog turned pale, thought about the flack he would catch, thought about that raise he would never get, thought about being sent to the further hinterlands where supervisors are forgotten about until retirement time, and did what any sensible man would do. He used what tools he had in the first aid kit and spent some two hours pulling splinters with tweezers and coating the wounds with antiseptic.

All this time he downplayed the seriousness of my injuries. "Just a few scratches. Man don't go running to a doctor over something like this. Ha! Ha! You're gonna be good as new."

It worked. No infection, no permanent injuries.

Boog would soon lose his located job at Clinton, but it wasn't because of this incident.

What griped me was he didn't buy me a new shirt.

FOUR

BIG CITY LINEMAN

Listen my children, and you shall hear
Of the midnight ride of Paul Revere....

My midnight ride wasn't quite as dramatic. Still, it was exciting enough.

While living at Clinton, our second son was born. Since our family doctor lived in Guthrie, when the time drew near Dixie stayed with family. Of course Larry arrived a month or so after the doctor's projected date.

In a desperate attempt to hurry him along, Dixie swallowed a large dose of castor oil. This made her so sick that she hoped they could save the baby because she knew beyond a doubt that she was going to die.

We didn't have a phone so our landlady consented to take the call. Like most babies, Larry was determined to inconvenience as many people as possible by keeping them awake most of the night. So, he waited until nearly midnight to begin his journey into this world.

It was near the midnight hour of February 11, 1952 when the landlady knocked on my door and told me the call had come. I headed for the Guthrie hospital, driving that old Ford faster than was good for it or me.

As I arrived, they were wheeling Dixie back to her room. Larry was the only one of our children born when I was away. Not that my being there helped the others.

After returning my family to Clinton, we decided the time had come to move. Service had been provided for rural folks who wanted a telephone. The Company was soon to dissolve the Clinton District.

Word was that this territory would be annexed into Enid. Boog would be an unlocated construction foreman and Slick would take his job. In the spring of '52, these rumors became reality.

Glenn, a victim of downsizing, was busted from second-line to first, a bitter pill for any man to swallow, but especially for one with Mr. Wilkerson's status in his home town. There were those who thought he would be a tyrannical leader. I didn't doubt them one bit.

It is difficult to predict how any individual will react to adversity. Those linemen who stayed vowed to the man that Glenn was the best boss they'd ever had. So much for crystal ball gazing.

Our other options were to quit the company, or transfer to the city I had earlier rejected, the one place I definitely didn't want to work or live, that Sodom and Gomorrah suggested by Mr. Fuller in the beginning and promptly rejected, and that was Oklahoma City.

Lloyd decided to take another exit route. He'd had his fill of pole climbing. Having done body work on cars before hiring on with Ma Bell, he decided to return to this vocation. Here was another good employee who checked out many years short of pension eligibility.

That same spring of 1952, a located lineman was needed in Oklahoma City. I placed a bid through the Union and, to my surprise, within weeks, found myself heading towards the big city. There was more than a little difference between the two jobs.

We rented a small apartment and moved what few possessions we owned. I was told to report to the work center at 7 East 6th. Compared to what I'd left behind, this was like stepping into a beehive. There were construction crews and cable splicers, along with telephone installers, repairmen, mobile phone installation, motor vehicle maintenance, and a flock of elite folks who installed and maintained large business systems. One look was enough to make a man long for those good old days in the country.

My first morning, while waiting for the eight o'clock starting time, there were snickers and giggles as word circulated among the men as to who my new boss would be. Listening to them caused me to have doubts about leaving Clinton.

My new boss and Boog were as different as sugar and salt. They were both white but there the resemblance ended. Old Boog, the extrovert, always laughing, was, off the job anyway, much like one of the boys. This boss was so introverted that no person really got to know him. He rarely spoke and his smiles were rare. Behind his back, he was often called "Smiley".

Trouble seemed to follow the old boy, even more than it does most folks. A story I heard, whether or not true, I don't know, went like this: he had once stopped a man on a bulldozer who was getting dangerously close to one of our underground cables. Smiley lit into said man with words not spoken around preachers or ladies in those days. The driver, being neither of the aforementioned, bailed off of his machine and, so the story goes, worked my boss over with a large Crescent wrench.

It was said that Smiley held the world's record for most fights started without a single victory to his credit. Winter or summer, clear or cloudy, indoors or out, he always wore dark glasses. To conceal bruised and maybe blackened eyes, it was maliciously rumored.

Having escaped reporting to Slick, I found myself reporting to a peculiar man.

At Clinton, I had been top man in seniority; now I was last. Barely. My fellow crew members were amiable and made me feel welcome. Most construction crews consisted of a foreman and four linemen. Once I was added, we all stayed together for several years.

Vernon Shaw was the largest man on the crew and I was always glad that he had a good disposition. Vernon wasn't fat, just big as some folks are. Like John Wayne, for example. For a big man he wasn't clumsy. In fact, he was better at walking wood than were most folks. He was a pleasure to be around, the kind of man you would want at your side if trouble came. I worked with Vernon for several years and then he left the crew because of an event we'll discuss further in another chapter.

Bill Kincaid was a tall, lanky man with a dry wit, who was good for a laugh or two every day. He wasn't known for speed; however, his methodical tenacity along with a tendency to make very move count, often allowed him to complete a job while those of us who valued speed above all else were kept busy redoing what we had done poorly in the first place.

Bill eventually moved to Commercial and worked his way up to District level, a remarkable accomplishment for a man who earned his college degree while working full time and attending night classes.

From that first day, when we rode together in the back of an old construction truck, Vernon, Bill and me seemed to hit it off just fine.

Yes, the back of the truck. No more five-man cabs, not for big city linemen. Since we didn't normally drive across country, we rated only the cheaper models.

We rode in the back, sitting on a cover over the winch-line, on tool boxes, or whatever else was handy. During warm weather the rear was open so you could tell where you had been. During cold spells, we fastened a tarp over this pneumonia hole. For several years, that is the way I travelled from job to job--seeing where we had been but never where we were going.

Only the driver and foreman enjoyed the relative comfort of the cab. The truck driver was Jay Worden, a few years older than me, a Navy veteran, and when need be, he would stand up to Smiley.

Smiley had his own way of communicating, which was usually poorly. When we headed out each morning, only he knew our destination. For some unknown reason, he delighted in keeping this secret from his driver.

At the beginning of each day, we would climb on and he would point to an exit. Since we only made right turns into and out of the work center, this pointing gave some clue as to our eventual destination. Smiley's habit was to wait until we were near an intersection and jab a thumb in the direction we were to turn. This little game would continue until we arrived at the job site.

One day we were on our way to some secret destination and Smiley forgot to point. We were upon the intersection when he grabbed the steering wheel and pulled across traffic and into a left turn, throwing us who rode in the back hither and thither, banging off the sides of the truck and dodging swinging tools that were potentially lethal.

There was screeching of tires and unkind words from other drivers.

Once Jay had gained control, he pulled to the curb, got out, came around to Smiley's side of the truck, and said, "That's it! You want to drive, go ahead. You don't, then from now on you tell me where we're going before we leave the yard."

The back of the truck had a sliding canvas top that we kept open during fair weather. Therefore, we in the back were able to see some of and hear all of this exchange. Smiley mumbled an address and soon we were on our way.

It was not unusual for Smiley to sit in the truck while we worked, especially in cold weather. He would stay inside with the heater on, nodding off now and then, awakening occasionally to either roll the window down or climb out, if need be, to issue orders. I suppose this was his prerogative, but most of the foremen didn't seem to operate in this manner.

Not only was my boss different from Boog, but the work differed also. Besides climbing poles, we often worked in office buildings and in smelly manholes. The work was hard. We had no hydraulic equipment to use in digging holes and sitting poles. Holes were dug by hand although we did have a makeshift truck derrick to use in setting them in the ground. It seems as though every week, and often every day, we would be doing something new. Whatever else they might have been, the six years I spent as a lineman in Oklahoma City were not boring.

There was a hazard to pole climbing I had never once expected. It seems as though a man's pockets are designed to hold too much and yet never enough. One item I carried in a front pocket was a honing stone, used to keep gaffs in tiptop shape. Being a cigarette smoker at the time, in the same pocket I carried kitchen matches.

While climbing a pole, I thought I smelled something burning. It was me. There was a hot spot on my leg. From some fifteen feet up, it took only two steps to hit the ground. A match head had rubbed the stone, ignited, and then fired others. After I had extinguished the blaze and was examining my ruined pocket and burned skin, there were some folks who thought this was funny. Even old Smiley chuckled.

One summer when Smiley was on vacation we worked for Joe Bobo. Joe was a man who loved to eat, attested to by a rotund figure. Anyway, we were doing maintenance work on one of the toll-lines. Joe had me attach clips to wires so he could have a talking circuit. On the ground he had a test telephone complete with hand crank, used to signal the operator.

Joe cranked and cranked. Into the mouthpiece, he shouted over and over, "Hello, operator! Hello!" When finally she answered, in his slow, deep voice, he said, "Operator, this is Joe Bobo, the telephone man."

That's how we viewed Joe--as *the* telephone man.

In contrast to Smiley, Joe was talkative and outgoing. You had the feeling that the old boy really cared for your welfare. There wasn't a man on the crew who wouldn't bust his butt for Joe.

Another summer Smiley was loaned for a couple of months on some special assignment. When we heard that Joe would be our temporary boss, we were elated.

A job we worked on was adding four wires to a toll-line east of Oklahoma City. This was an agreeable enough job. Walking wood in hot weather sure enough beat working beneath the ground in smelly manholes.

Joe was not an early riser. Each morning, about eight, some forty-five minutes or an hour after the other foremen had arrived, Joe would waddle into the yard and say, "Boys, get 'er ready to go. I gotta go to the toilet and call the engineer and then I'll be with you."

Somewhere along the way, he would have Jay pull into a cafe parking lot. We'd all get out and Joe would say, "Boys, I gotta have a little breakfast before we go to work." The rest of us would drink coffee while Joe ate and then on we would go. Once we reached the job site, we would work like men driven with a mission.

Come noon, we would gather under a shade tree for our thirty minute lunch break. We carried an old deck of cards on the truck. While eating, we would have a game of penny-ante poker. Joe would always join in.

If he was winning, our half-hour lunch would usually stretch to forty-five minutes or an hour. Then we would work our tails off. Often it would be quitting time or past before we would climb down off the poles and begin the long trip back to Oklahoma City.

Even though we would arrive at the work center a half hour or sometimes even an hour late, not one of us ever complained because Joe didn't pay overtime. When a radical Union member from another crew suggested rather strongly that we file a grievance, we circled him like wolves around a crippled calf. After a heated verbal exchange, he backed off.

While working as a lineman, we bought a house in Del City. We were living there when the remainder of our family was born. Delmar arrived first, then Patti. At long last, we had a daughter. Of course having three older brothers, she was a tom-boy.

While living in Del City Dixie learned to drive. I insisted that she do so. She would shudder at the thought of shifting gears. Once I had ridden with her, I shuddered at the thought of her shifting gears.

So, I bought a used, '55 Chevy with an automatic transmission. That took care of that.

Although changes within the company were rare, they did occur. Most of the old-timers hoped to preserve the status quo. However, new blood gradually filled some of the higher management positions. These folks began looking for innovative ways to run the company. One idea that became reality was the formation of two-man construction crews.

This was a bitter pill for many of the old hands to swallow. Traditionally the foreman had spent the day with his crew, jealously guarding his position. He would read a work print and then assign a certain job function to each man. It was he who made the daily time reports, charged off the major items of material that had to be accounted for, and made sure each man was paid. With two-man crews, all this fell by the wayside. Linemen took over these duties.

I was sent to another work center in Oklahoma City where I was paired with an older lineman, Kenneth deCordova, known as "Dee". We hit it off just fine.

Dee's head was overflowing with knowledge. He probably knew more about telephone construction work than any man I've ever known. He'd been a construction foreman in times past, but had become a victim of reorganization, a downsizing of forces.

I preferred pole work to being on the ground. Dee's preference was just opposite so when only one man was needed aloft, that was me. This then was an arrangement that suited us both.

Some of our work was done at the Tinker Field Air Force Base. We had passes, just like those carried by regular Tinker employees. One day a Company Engineer asked us to check inside a manhole on base to see what ducts were vacant. We found the manhole, pulled the lid, and climbed down the ladder. We were making notes when a voice from street level ordered us out. The man sounded as though he meant business.

Somewhat apprehensive, we climbed to the top. Air police had us surrounded, rifles at the ready. We were loaded into a Jeep and driven to headquarters.

We had wandered into SAC, the Strategic Air Command area. For this, you *must* have a special pass. We had ignored the signs. The warnings could not apply to a couple of hardheaded linemen. It took an hour or two and a number of phone calls before these good folks were convinced we weren't Russian spies.

My last six months or so in construction were with Jimmy Jobe. Jimmy and I both liked to climb poles and we would often compete to see who could complete a job the quickest. In retrospect we might have hurried a little too much but fortunately neither of us had an accident during this time.

My days as a lineman were drawing to an end. It was time to move on. But first, a brief interlude to talk about storms.

FIVE

THE FIRST STORM STORY

In the early days of telephony, all wires were bare and exposed to the elements. They would swing from pole to pole or from tree to tree or from any tall object to another tall object, and were tied to insulators. Years would pass before individual wires were insulated by paper and wrapped in a lead sheath cable, this being the next giant step in outside plant technology.

Originally all lines were whoop and holler, one wire to ground. When electrical wires were near telephone lines, the two proved to be incompatible. Telephone engineers invented the metallic circuit, meaning it now required two wires to service a customer, unless theirs was a party line, and then this same pair of wires could accommodate eight, ten, twelve families or more.

All wires had to be attached to insulators, usually ten to the crossarm. The custom was to set tall poles and build down from the top. Need another pair of wires, you string 'em. Need another crossarm, you add it and then string wires. Need even taller poles, you change them out. Some of the photos of those early-day scenes would cause a modern day repairman to throw down his climbers and run away, screaming as though possessed.

In downtown New York, in those infant days of the Bell System, poles were often some ninety feet tall, holding thirty crossarms spaced two feet apart and a total of three hundred wires.

Wouldn't you know it--in 1888, along came a blizzard. I mean a rip-snorting, straight from the north pole, via Canada, ice forming, wind whipping, open-wire-line destroyer. By the time it had blown itself out, wires had pulled apart, many crossarms had snapped and broken poles were decorating the streets of New York City.

A cry echoed throughout the land and Bell engineers decided to get serious and find a way to protect their wires from the elements. They did it! Lead-covered cables were placed in underground ducts. Tall poles and wire lines began to disappear from downtown areas.

However, through the fifties, into the sixties, and even into the seventies, this same combination of poles, crossarms and exposed wire ran from town to town, linking city to city, so that with the help of a female operator one could talk to folks in far places.

Long before women manned the boards, boys were telephone operators. Young men, some reports suggest they spent more time goosing one another and hustling beleaguered customers than they did in placing calls.

In an unprecedented move Ma Bell took women out of the kitchen and sat them at the switchboards. Purists from all over the country began shaking their heads. Women did *not* belong in the work place!

What these ladies did was get the job done in a more efficient manner than anyone could have imagined. They were businesslike, well trained and soon the *voice with a smile* was what folks expected when they reached an operator.

No thought was given to bringing boys or men back to the boards until Uncle Sam forced Affirmative Action upon the old gal. Thus, men next appeared on the scene when Uncle flexed his muscle and efficiency be damned! Ma had to use them whether she wanted to or not. It troubled Sam to see a company running a smooth operation. Made him look bad.

No matter who ran the boards, storms kept on hitting. Since cross-country lines were one of Ma's top moneymakers; she hated to see them a tangled mess upon the ground. Whenever that happened, it was pull out all the stops. Call in as many troops as necessary to get the job done in as short a time as possible.

For every minute a line was down, not only was money being lost but public relations were taking it on the chin as well. Even so, because of the enormity of the catastrophe, it would often take days or even weeks to restore service.

Ice and wind weren't the only weapons nature used to wreck telephone lines. Now and then she would throw out a deadly tornado. Or a hurricane. Tornado damage is smaller in scope and more quickly repaired. Hurricanes can be devastating, but from a workman's viewpoint they are more enjoyable to follow than are most storms. Once the fierce winds have passed and the torrential rains have ceased, the weather is usually quite nice. As for an ice storm, one will freeze your buns off, believe me.

My first ice-break was November 1, 1951, nearly three months after becoming a member of Ma's family. As ice breaks go, it was only a baby. I was working out of Clinton, for Boog, and of course Line-3, the big toll-line, was running across country east to west, the most vulnerable direction for damage when the rain's a freezin' and the wind's a howlin'. Scarcely a wire snapped on this line.

What we worked on primarily were open-wire circuits serving local customers. For three days we scurried about, sagging thin wires and splicing them together. Main thing I learned was how to climb a slick pole.

Depending on how the wind blows, one-half to some two-thirds of a pole will be coated with ice. Now old Boog himself had taught me to climb a pole on the high side. Always, he'd said. Emphatically.

If a pole is leaning an inch to the south, climb it on the north. If it's leaning a little to the east, go up the west side. So what right did he have to come screaming at me when, encountered with my first icy pole ever, I decided that the lean was to the north and so I would do my climbing on the south.

We all know Boog wasn't bashful. Instructions poured from his mouth. He figured any numskull should be smart enough to stab a climber through ice. Thus, as you ascend or descend, your hands will clutch wood instead of slipping and sliding. Boog's explanation, if not tactful, was certainly easily understood.

So much for my first ice break.

While working for Smiley, we were sent north to Wichita, Kansas to lend a hand to those folks since ice and wind had played havoc with their wires and poles. It was quitting time one cold, snowy day when old boss man told us to hurry home, pack a suitcase, scoot on back, and load up. My fellow crew members were still Shaw, Kincaid and Worden.

By the time we returned to the storeroom, it was dark. Jay was the truck driver so he and Smiley climbed into the cab where it might have been a degree or two warmer than outdoors. The rest of us climbed in back where the temperature was the same as that outdoors.

A tarpaulin covered the rear but we shivered and shook. We froze our tails off until late, very late that night we found Wichita and fell into beds awaiting us at the English Village Motor Lodge.

Although this ice break was of a much larger scale than what I had experienced while working for Boog, it was still small as ice breaks go. Much of our time was spent working on little rural lines. Mother Nature did not cooperate. Freezing rain and sharp pellets of sleet made conditions cold, slick and hazardous.

Since we didn't visit socially back home this was my first time to be around Smiley, other than on the job. One fact I soon learned was that as soon as we quit work each night, he would head for his motel room and hit the bottle. It seems as though it didn't take long for the contents of said bottle to hit right back.

Some two miles from the motel was the cafe where we took all our meals. Having no other means of transportation, we travelled to and fro in the truck.

One night we drove in from the country after dark and had to wait some thirty minutes or so before Smiley emerged from his room, ready to head to supper. He was tipsy and tottering and insisting that he drive to the cafe.

Over Jay's protests, Smiley slid in behind the wheel. All we could do was climb aboard. Either that or go to bed hungry. Jay rode up front with the old man.

The truck roared down the boulevard. Smiley saw the red light ahead and something in his whiskey soaked brain told him to stop. His foot slammed down on the brake pedal. In back, we were thrown around among ropes and blocks and chain hoists and all the other lethal equipment that hung on side rails. This happened twice before we reached the cafe.

Inside, we sat together in a booth, Smiley on the outside, drunker'n a sailor on his first shore leave in six months. When the pretty little waitress with a smile on her face moved in close to take our order, for reasons known only to Smiley and God (I'm not sure that even God understood Smiley), the old man decided it was time to reach out and touch someone.

Before we could stop him from doing so, he ran his hand under her dress, touching places that nice girls don't appreciate being touched other than by someone in her life who is intimate. She screamed. Fire in his eyes, the proprietor ran to her rescue. The tearful waitress cried out a story that none of us could contradict.

We escorted Smiley out of there in a hurry and had to change our eating place that night and from that day on. Jay drove. No ifs, ands, or buts about it.

My next Kansas ice break, I was working with Dee on a two-man crew. At the time we were reporting to Bill Countryman, a plump, jolly man with a country sense of humor. We didn't work at Wichita, although we did have to wait there for the better part of a day while snow plows cleared the roads west.

We then travelled to the western part of the state. Jackrabbit country. Trees were as scarce as upper teeth in a cow. There wasn't a blamed thing to block that icy north wind.

The storm had been devastating. Everywhere we looked fallen telephone lines littered the ground. Not only their rural wires, but those that ran across country, the toll-lines. The ground was covered with a thick layer of ice and on top of that lay a thicker layer of snow. Freezing winds, buffeted by nothing larger than a fence post, howled out of the north.

We spent days tromping through snow that in places was nearly knee-deep, salvaging what we could of the old wire and splicing in new when necessary. Poles were set; crossarms hung. When we were ready to make the line permanent, sag the wires to their proper tension and splice them together, I was the designated pole man.

Up I went and then, on a handline, pulled up blocks for stretching wires, sleeves for splicing, and a sleeve roller so they wouldn't pull apart. With frozen feet, I tromped up the pole. Wet clouds kissed the earth. A stiff, Arctic wind penetrated my clothing, searching for the tiniest opening. Eyes watered. Much of the work had to be done without gloves. Cold fingers fumbled in the bag for tiny sleeves used to splice wires together.

Wires were fanned out from the east and from the west. I pulled them up off the ground, sagged them to the proper tension, and spliced them together. Much of the time I had nothing to do but wait and shiver with my back turned to the wind. Even a horse is smart enough to do that.

I spent from daylight 'til dark upon that desolate pole, coming down only long enough to eat lunch. I believe this was the most miserable day I have ever known.

Within days Mother Nature decided to change her weather once again. The wind subsided; the sun poured its brightness upon us. We wore dark glasses to protect our eyes from the glare. Many of the men suffered from a sunburned face.

One man, a construction foreman from Texas, fared the worse. His face puffed up until you could scarcely see his eyes. He was not only miserable but quite sick. He was sent home for medical treatment. I suppose that accident could have been avoided also.

One of our more memorable ice breaks was at Jonesboro, Arkansas. Dee drove over while I enjoyed the view. Dee reminds me that I drove the last leg of the journey while he studied a map with a flashlight, trying to find Jonesboro. He did! We arrived, late at night.

Jonesboro isn't typical ice-break country. It is north of Memphis, not far from the Mississippi River. When we reached town, there was neither ice nor snow to be seen. However, it was pouring rain and continued to do so all that night and into the next day.

The company had a warehouse where we not only stocked our truck with supplies but selected hip boots and rainsuits. I can testify from personal experience that climbing a pole wearing hip boots is comparable to climbing one while barefoot.

According to the natives, what brought their open-wire lines down was freezing rain, but no wind. Ice, they said, grew to a diameter of four inches and wires snapped from dead weight. When enough wires snap in one span, the pull from the next can snap a pole. And then another. Or the end of a crossarm. Their toll-line was in one heck of a mess.

The line ran parallel with a railroad. Many of the new poles were set by contractors who stayed ahead of us. How they managed to get some of them into the ground, I don't know. It wasn't unusual for us to have to wade through two feet of water before climbing.

All rules have to be bent. There was no way we could possibly strap on climbers one step from the pole. The revenue from toll-lines was considerable and the continued cost of room and board for construction crews was enormous. Yeah, we bent the rules. You don't sit around waiting for the water to subside even though by then the sun was shining. Once the rain stopped, the weather at Jonesboro was near perfect.

These were my more memorable ice breaks to work on as a lineman. We did chase after a couple of tornadoes. One, near Meeker, Oklahoma, followed a wire line down the railroad tracks for a couple of miles.

Another hit our line between Edmond and Guthrie. A little twister roared through on a Friday afternoon, not two miles from where I had been born and raised and scarcely more than that from where my folks lived at the time. I had barely arrived home from work when Smiley phoned and invited me to the party. The mistake I made was in not eating supper. Also, I should have packed a midnight lunch. Maybe breakfast.

The tornado had cut across the line instead of tearing along parallel. Consequently, damage was limited to six broken poles and slack wires for a ways beyond.

We worked all through the night, digging holes by hand and setting new poles. Then we spent until mid-afternoon Saturday replacing the wires, pulling them to the proper tension, and tying them to new insulators. Not only had we not eaten supper the evening before but we didn't get a midnight snack. No breakfast nor lunch on Saturday. As my stomach complained, I decided that Smiley was sadistic.

What made it worse, there we were only two miles or so from one of the world's best cooks, my mom, and there I was, nearly starved to death. When finally we finished, we stopped at Edmond and ate until we could eat no more. I then hurried home, jumped into bed and didn't awaken until some time Sunday afternoon.

Oh, to be young again. Looking back, those were in many ways the good old days. An old neighbor once said that the best thing about the good old days is that they are gone forever. Maybe so. But so are toll-lines.

Open-wire telephone lines are becoming scarce. Copper wire has been recycled. Poles have been sold to contractors or to farmers. The insulators, those green, blue or white glass fittings that once decorated poles, have gone the way of the steam engine.

The proper name for those used on toll-lines was "double petticoat." When you study one, it's evident why. Most are long since gone. They were used to help fill the holes when poles were pulled. Then they were considered to be only worthless glass. Now they are treasured by antique buffs.

Yeah, them good old days.

SIX

SHORT SLEEVES/LOW-CUT SHOES

Thomas Watson, a man of many talents, not only contributed to the invention of the telephone, but so far as we know was the first person to ever install one of the new gadgets. As a technician, he was clearly frustrated when words spoken over his phone didn't sound all that clear over a long distance of wire. So, he did what any reasonable man would do: he consulted a spirit medium. Word was he even tried a little table tipping.

Mr. Watson all but worshipped old A. G. Bell. One of Bell's habits that really impressed Thomas was that the inventor's primary table-tool was a fork instead of a knife. Imagine! Just how sophisticated can you get anyway?

Working for old Ma, you might feel frustrated, belittled, persecuted, or set upon. You might be hurried, harried, enervated, and perhaps even bewildered, but seldom were you bored. There were always new people to meet, new bosses, new coworkers, new locales, and now and then, perhaps even a new job title.

Since she had a contract with the Communication Workers of America, Ma was forced to advertise all craft job vacancies. Any man with a craftsman's title (only men held such titles back in those days), could bid on any job open within the Division where he was working. If you coveted an advertised job, then you mailed all of your pertinent information to the Union. If you were a member in good standing and had not had a successful bid within the past year, they would then throw your name into the hat. All legitimate bids were forwarded to the Company.

According to Company propaganda, they considered both seniority and qualifications, supposedly giving each of these factors equal weight. The Union insisted that all jobs be awarded on seniority alone.

The Company was able to save face by saying they considered qualifications. With few exceptions, however, seniority was the primary criteria. The burden of proof was not on the Union to prove an applicant qualified, but on the Company to prove that he was not.

The Union's argument was: "If he isn't worth a hoot, why did you hire him and keep him on the payroll for all these years?" Building a case against an individual in an attempt to keep him from being the chosen one was time consuming and could even expose management to charges of improper training. To heck with that hassle! The Company would usually take the man with the most seniority and be done with it.

After seven years as a lineman, a bid caught my attention. Four additional installers were needed in Oklahoma City. Although there had been a time when the demand for these folks was such that they were hired off the street, not so in the late fifties. Between the end of World War II and the time I was hired, the Company nearly exploded from the influx of new-hires. At the time

I didn't have enough seniority to realistically hope for anything more than a lineman's job.

What the heck? I sent in a bid. Danged if I wasn't fourth in seniority of those who did. Since the Company did not go to the trouble of proving me incompetent or incapable, I was soon out of the manholes or all day pole climbing and thrown in among the lesser prima donnas, those who were a step down from the PBX installers.

Linemen were still required to wear long-sleeved shirts, hats or caps, and safety glasses. Nothing had changed in this regard since that first day I went to work for Boog at Sayre. An installer could wear short-sleeved shirts, low-cut shoes, go bare-headed if he so desired, and was not required to wear safety glasses other then when performing certain work operations. Of course if we were up a pole, all these items were supposed to be available on the truck and put on before going aloft.

Yeah, you bet.

Anyway, I was now an installer of telephones. I reported to Tommy Morgan.

Many in management, especially those higher up the ladder than first-line, feel as though barriers have to stand between a craftsman and his boss. A supervisor should goad his subordinates, jab them verbally, belittle their accomplishments, push, drive, shout, shove, and be overly aggressive, never for one moment allowing the ungrateful creature to forget his place. Some tried this approach. Usually it backfired. My new boss would have nothing to do with such a method.

Tommy's way produced results. We gave our all. And more. Instead of taking advantage of his good nature, we were careful not to let Tommy down. Later, as an installation foreman, I especially felt his influence.

I stepped into a new world. In the telephone installation department, a third of us worked any given Saturday and were scheduled a day off during the week. For the first time ever, I was faced with the pressure of keeping customer appointments and installing their telephone when they wanted it.

A customer had called the Business Office either for new service or for some change in existing. The Rep. had promised the customer an installer would do the job on a certain day. If the customer asked, they could get a promise of a morning or afternoon appointment. If a customer insisted, they could specify a particular two-hour time period. Many asked; some insisted.

When dealing with the public, customer service is the byword. To the extent possible, we had to give the customer what they wanted when they wanted it.

Usually I worked alone. There were occasions when I would help another or he would help me. Late one evening I called for assistance. I had a phone to install inside a large warehouse where an office with glass windows occupied one corner. Along came a buddy.

Inside the building, wire had to be run above plate glass windows. At that time, I smoked a pipe. We had the extension ladder run up some twenty feet. I was on the ladder, puffing on my pipe and working away. Since the ladder was sitting on concrete, the bottom had to be held in place to prevent it skidding.

The wire hung up under something or other. My friend turned away to clear the snag. The ladder began sliding down the wall. He ran back. I came close to plunging through the glass before he stopped the slide. Only after climbing down and taking stock did I realize that I had bitten my pipe stem in two.

For several years, my route lay next to that of a friend I'd worked with in construction. While on a two man crew, Al Hester and I had reported to the same boss, Luther Steed. Now, as installers, it became our habit to get together each morning at the coffee shop, have a cup and compare service orders. We made it a point to always help one another before checking with the dispatcher to see if anyone else needed help.

There were times when it would take more than eight hours to work our orders. When you offer a choice of date and time, the work load naturally fluctuates. We often worked late. Climb a pole with the aid of a flash-light. Sometimes a penlight, held in your mouth. People might think we were crazy but they got their phone.

Since we often entered fenced yards to reach a pole, we had to be on the lookout for dogs. Late one summer evening, I was installing a phone in a working class neighborhood. It was supper time for most folks. The pole I needed to reach was inside a yard next door to my customer.

I scanned the yard for dogs. None were visible. The pole was stepped, so I didn't have to wear climbers. I had passed through the gate and was near the pole when I heard a noise from behind.

This big German shepherd had been inside the garage and had evidently awakened to the knowledge that an intruder had invaded his domain. He charged out, intent on attacking and destroying. Although I usually detested stepped poles, I thanked God for this one. I was up and out of reach before he could sink teeth into flesh.

Problem was, how would I get down? Old dog showed no inclination to be friendly nor to retreat to his garage haven. No help in sight. Finally I traced the line going to the canine owner's house, followed it to inside the terminal, connected my test set, then dialed the ring-back code. I switched to monitor and listened. When the man answered, I flipped off the monitor.

"Sir," I said, "I'm on a pole in your back yard. Will you please come and get your dog?"

He thought it was funny. Some sense of humor. I felt like a treed opossum.

Almost all the phones we installed were black. Unless the customer lived far away from their serving central office, we were required to connect the old 302 model. These were the heavy, nearly square instruments with the rounded handset that were continually installed, disconnected, cleaned and then installed for another customer.

These old phones were awkward and ugly, but they were as durable as dirt. When the storeroom ran out, we would work our disconnects first, clean the removals, put new cords or a handset on if needed, perhaps change out the ringer, and then we would install it for the next customer.

By this time the 500 series telephone was available not only in black but in other colors as well. They were lighter weight and even comparatively modernistic. Any available color could be installed for any customer willing to pay the extra ten bucks.

We would carry in the old, rebuilt, ugly, dingy 302. The customer would take one look, come close to going into shock and demand one of them new type phones. The time had come to make our sales pitch even though the Business Office Rep. had hammered away at the poor soul in an attempt to persuade him/her to buy not only color phones but extensions and bells and push buttons and buzzers and anything else they could locate in their sales book.

Now we had one last shot at the reluctant soul. If she refused to pay the extra ten bucks for a paint job, she would soon be dialing on a relic.

Equipment sales were very important to the Company. We were under pressure to sell extensions, color phones and the new Princess. Of course the customer had already suffered through the spiel of the fast talking girl from the Business Office who was under extreme pressure to sell and would try every trick and gimmick known to woman. Some would even pad the order. Yes, they most certainly would.

When we arrived on premise and carried in all those boxes, the customer would more than likely tear into the po little installer for being such a nincompoop.

Invariably she (Our contact was usually a woman) would say, "I told her (Our Business Office Rep was always a woman) that I was not going to pay for all them telephones. Listen to me, young man, I want one black telephone and I want it right there!" A finger would be pointed at the location.

"Yes, ma'am."

So I would tote the phones back to my truck and scrounge around for a black telephone, all the while muttering unkind words towards the girl in the Business Office who had talked to this customer. For them though it was survival. Their job depended on sales.

What they'd hoped for was that the customer would change her mind by the time we arrived. If not, since we had phones in hand, it was hoped that we would convince said customer to keep at least part of them, thereby giving the salesgirl credit.

We would complain about their shenanigans; they would in turn complain that we were only too lazy to install the additional equipment. There were often feuds among the various departments whose goals clashed.

It was while I was installing phones that Dixie and I decided to move our family to the country. In 1960, we took the plunge and began to build our own home. By early summer the footing and stem-wall were poured.

I took all of my vacation and hired help to frame the structure. Our kids ranged in age from five to nine and of course were anxious to contribute. For them being in the country other than on weekends was a new experience.

There were oak trees near the house. After being told some ten thousand times to stay out of the way, the kids decided to gather up scrap lumber and build a treehouse. None of us thought much about it when Delmar began swinging a hatchet at something on the ground. Not until a neighbor discovered it wasn't a board Delmar was chopping. Instead, he was swinging away at a coiled rattlesnake!

By God's own mercy, our son was not bitten. This was one of four rattlers we killed that summer, all within a few feet of the house. Evidently we had broken ground right in the middle of their homeland. They weren't the forgiving sort.

When the time came to move, Dixie was reluctant.

An interesting job assignment I had during my years in installation was upgrading pay phones in our district. This meant changing them out to a newer model. To do so, I had to be bonded since I would be collecting money from the phones and installing money boxes in the new ones.

The Coin Office was located at the work center where I reported. They were responsible for seeing to it that all money was collected and counted. There were times when an installer or repairman would be working on a pay phone and find money that hadn't gone on down into the box. If so, we were to drop it in. To keep us honest, folks from Security, with the help of Coin Office personnel, would sometimes salt the phones, plant coins coated with a chemical that would only show under a special light. After a craftsman had visited that phone, they would check the money box. If the money wasn't there, they would then check the craftsman and usually find the money. Since this was an act of dishonesty, it was cause for termination.

I would go into the coin office each morning and hand Midge Sullivan a list of the phones I expected to change out that day. She would give me keys to these phones along with empty money boxes to install in the new. Midge wasn't the supervisor over the coin office at the time, but she knew how to take charge.

One phone I replaced was inside a wooden booth, inside a dark tavern. I removed the outer door and was in for a surprise. No money box was inside. Instead, there was a pile of loose change. I was a little apprehensive, as I wondered if this was a test of my honesty.

I immediately phoned the coin office where Midge put me through to the manager, Mr. Hilton. I told him what I'd found and he said that there was a man who went from city to city and was able to pick the locks, remove the coin boxes and lock the outer plate back on. This was a remarkable feat since they were all keyed different. I was to count the money and place it in an empty coin box. And so I did.

During the five years I installed telephones, I saw the Black Community from an entirely new perspective. By definition I was probably a bigot, racist, and redneck. I had certainly been programmed that way.

As a child I had seen the separate restrooms at the train depot, as well as separate water fountains almost anywhere a business provided public drinking. We had our grade school; they had their own.

There were times when my grandpa would hire a black man to do field work. It was the custom in those days that come noon you fed the workman, no matter what his color. My grandma would fix him a plate and he would sit alone outdoors while those of us who had been born white, by the luck of the draw, and nothing more, bellied up to the table.

I didn't hate blacks or wish them harm. The plain truth was I didn't understand them. Now I found myself going into their homes and installing telephones. It was enlightening. This was before so many of their young people seemed to be divided into gangs.

Some white folks thought of blacks collectively as being lazy, shiftless, and no good. What I found was that most were friendly, some were suspicious, while others were clearly frustrated. They were not lazy; they were simply shut out of the job market.

I saw women with four, five, as many as eight children, surviving in pitiful housing. Perhaps she'd once had a husband. Unable to provide for his family, he had left. His kids would then qualify for Aid to Dependent Children.

Our government, in all its wisdom, managed to wreck families and encouraged procreation. The more kids a woman had, the larger her check for ADC. Years before, the Indians had been confined to reservations. Now the Blacks had theirs. No, it wasn't called by such a name, but segregating them in the manner we did had a similar effect.

During this time there were often rumblings of militancy. One day I worked in a small but neat home installing a telephone for an elderly black couple. The old woman was working in the kitchen; the man was in the living room, stumbling drunk and still tipping the bottle.

"How come your Company don't have no black man out here installing my telephone?" he wanted to know.

I mumbled some vague answer. As I stapled inside wire, he continued the conversation and became adamant about the fact that the Telephone Company at the time had zero black installers. I was wanting to finish and get the heck out of there.

When I looked up, he had a pistol pointed in my direction. That can be scary. At a distance of six feet, even a drunk might get lucky. His voice grew louder as he continued lamenting the lack of black installers. I didn't know whether to wait, fight, or run.

I was saved from making the wrong decision when his wife rushed to my rescue. She entered the room wielding a long, sharp kitchen knife. The old man might have threatened me but he knew when he was outclassed. She chewed him up one side and down the other. As though she was dealing with a child, she took the gun and ordered him to sit in silence on the couch. He did. She remained in the living room until I was finished. I certainly was grateful to that old woman.

Still, I knew that the man had spoken the truth. Mr. Fuller would not, could not, hire a black man for any craft position. Nor could he employ a black woman to work in the Business Office or as an Operator. The best that any of them could hope for would be a menial job in house-service (sweeping, mopping, cleaning, waxing, etc.), or perhaps a job in motor vehicle maintenance.

That's the way it was. Not only with Ma Bell but with other companies as well.

A recent ad in TIME Magazine by AT&T says that Ruby Dee, a black woman, worked at a skilled job for Western Electric in 1942. It also claims that Lewis Latimer, a black man, at Alexander Graham Bell's request, drafted the plans for the first telephone. I'm quite sure that AT&T can back up their claim, but I'm equally as sure that during those times there were few blacks in responsible positions.

As for having a gun pulled on me by a drunk black man, I didn't report the incident to my boss nor did I call the cops. It seemed as though these folks had enough troubles without me adding to them.

Yes, I became disgusted with blacks because of their militancy, marches and sit-ins. However, if they had not forced the issue they would still be wondering why there were no black telephone installers.

One of the first African Americans to work in our storeroom was a man who had spent some twenty years working in house service. To be a good stockman, it is necessary to learn the terminology. Installers have their own name for most of the material. To them a "drop wire clamp" is a "P-clamp." A "drive-hook" is a "J-hook." Seldom is any item referred to by its catalog name.

This is enough to confuse anyone. Especially so when the stockman is an aging black man who feels intimidated. One day I went into the storeroom and found the poor man in tears. When I asked why, he said that some of the installers had been making fun of him, saying how dumb he was, and things like that.

He said, "Don, for twenty years I didn't have to learn anything new. I worked alone in empty offices or restrooms, doing the same thing over and over. Man, I didn't have to think, just work. Now they expect me to learn everything in a week. I can't do it."

I offered some worthless advice such as, "Hang in there," or "Keep your chin up," or something along that line. Still, I felt sorry for the man and regret for the biting verbal jabs flung in his direction by some of my coworkers.

After five years of installing telephones I got the one big break I had been waiting for. More on that later. First, more storms.

SEVEN

MORE STORMS

Once upon a time all hurricanes in these United States were routinely given a feminine name. Then someone noticed that they were often anticipated with dread, feared by everyone who happened to be in their path and more destructive than any means ever controlled by man, with the exception of nuclear power.

Some folks thought women had been getting the short end of the stick since the beginning of time. Heck, they weren't the only sex who exhibited many of these attributes. To their credit though, the ladies did play fair with us. Instead of giving male names to all the big hurricanes until the score was even, they agreed to begin anew, alternating between feminine and masculine.

For a chauvinist this was somewhat devastating. The arrival of the big wind is awaited with anticipation, filling us with fear and anxiety, as we wonder who she will favor and who she will not. She is fickle, vengeful, compulsive, and quite destructive. She roars through like a woman scorned: whipping, tearing, screaming and at the same time letting loose a billion-trillion teardrops, or so it seems. Small wonder that men hated to see their name on one of these storms.

In September of 1961 Mother Nature unleashed Carla upon the coast of Texas. It's possible that the Lone Star State deserved Mother's wrath but more probable is the premise that she just happened to be geographically located where collision was inevitable. Of course while Carla was throwing her terrible temper tantrum, she did damage in the millions to old Ma Bell.

Even before the big storm left the Gulf waters and slammed into the Texas coast, rumors were that some of Ma's Oklahoma children would be heading south to assist their Texas cousins. By the time she struck, we were ready to move. I was assigned to work as a repairman.

If you're planning to move an army en masse, you might as well harvest a little goodwill. Nothing is gained by slipping out of town in the middle of the night. While we prepared for the journey, the public relations people were making plans to milk the event for all it was worth.

We headed south out of Oklahoma City in a long convoy. Although I'd followed a number of storms by now, this was my first ever hurricane. Made me feel kinda proud being part of such an undertaking.

At Dallas, photographers, who were manning still and motion cameras, were on the overpasses, snapping pictures of our smiling faces. Efforts made by our Public Relations folks were now paying dividends. As I held my place in line, I'll admit that I sat up straight and had what I hoped was that *John Wayne at the Alamo* look on my face just in case I happened to be captured on film. I must not have been so captured since Hollywood didn't come a calling. Anyway, it was quite an experience for a country boy, albeit as part of a convoy.

There were eight of us from Oklahoma City who reported to a likable cuss, a thin and wiry little rascal.

You had the feeling that if confronted by the devil he would put up a good fight. He'd been in the Navy during World War II, a ship's officer, so the story goes, and you couldn't find a more patriotic man on either side of the Mississippi.

Jake's tag was Shaky. He was somewhat hyper with nervous eyes and a trembling hand. We figured that anyone who had spent years in the great war on a little ship in a big ocean had reason enough to quiver. So, although we did discreetly speak the tag-word only when our boss was not close enough to hear, no derogatory connotation was intended.

We travelled deep into Texas. Armadillos, some dead, some alive, were as thick as jackrabbits in Kansas. For several days we didn't go near the devastated area along the Coast. The planners of our expedition routed us around the major thrust of the storm so we were never directly confronted with Carla's terrible temper.

Our first stop was many miles from the Gulf, not far from Huntsville. Only a remnant of the storm had passed through this area, yet it was enough to cause significant damage. We stopped long enough to spend a couple of days repairing outside lines, so called *drop wires*, an insulated pair of wires covered by a single sheath, connecting a house to the main line. These wires were overhead; today, nearly all are buried beneath the ground.

I drove into the hills in search of a customer who had reported a dead telephone. As I climbed higher, the air became thick with the heavy but pleasant odor of pine. After much wandering and searching along mountain roads, I found my customers at the end of the trail. They were isolated from the world.

The house was old frame, well kept. Outdoors was reminiscent of a small zoo. Log fences surrounded primitive sheds. A donkey brayed while a beige colored cow stared in my direction and complacently chewed her cud.

Ducks quacked across the yard. A single goose spread his/her wings in a threatening gesture and hissed. Turkeys strutted while chickens scratched the ground for a living. Cur dogs greeted me with a little barking and a lot of tail wagging.

It was a step backward into time. There was no sign of any machine with a combustion engine. An iron-wheeled wagon was parked beneath a tree; I wondered if this was how they travelled to and from town when the need arose.

A dying Carla had roared through with enough impact left to snap a tree limb that, on its way down, broke the drop wire. I replaced the wire, connected it to the main line, and went indoors to test and see that service was restored

The old man and woman were inside. Ancient and cordial, they were pleasant to be around. They'd seen worse than Carla and survived. The woman had just pulled a coffee cake from her wood-stove oven. Fresh and hot, it was filled with raisins.

"Would you like a piece of cake, young man?"

How could I refuse? It smelled so good. "Yes ma'am, I would."

She cut a large piece, forked it onto a plate and seated me at the table. It tasted as good as it smelled.

The flies were quite bothersome and a goodly number had found their way into the house. I'd taken a bite when I noticed the old woman concentrate on one buzzing about. Like the tongue of an anteater, her hand lashed out. Fly in hand, she smashed the little critter between thumb and forefinger and let it drop to the floor. I swallowed hard, my eyes turned to the raisins, and I had to force myself to take another bite.

She repeated this feat several times before I took my leave. Not once did she miss. I've tried to duplicate her act, all to no avail. No coordination, I suppose.

The Powers To Be decided it was time for us to proceed to Freeport, a small city on the Gulf Coast. We rolled into town, amazed by the destructiveness of Carla. That old gal had packed a wallop.

Having been soaked with both rain and salt water, the city was slowly drying. A salty odor still permeated the air. The sun was now shining, as though Mother Nature was making amends for what she had wrought. Too late, Mother, the damage has been done. Sheets of plywood were being removed from windows of both homes and shops. Many of the natives seemed dazed. However, they were both resilient and resourceful.

Seagulls circled overhead, uttering raucous cries when several spied and fought over the same morsel of food.

A placid ocean stretched to the horizon. It was hard to imagine the waves that had tumbled ashore only days earlier.

The coast was packed with fishing and pleasure boats. Most were mangled, although a few were barely in working condition. We checked into a motel, found an eating place and discovered that some of us would be working at the Dow Chemical Plant.

The damage there was not so much from wind knocking down lines but from waves of salt water, belched by Carla. Ocean water does not mix well with telephones and cable and associated equipment. We started through the plant, replacing nearly every piece.

We couldn't help but feel sorry for the natives, many of whom had endured more than a person should be expected to endure. Still, I decided that if a man has a choice of storm jobs, working hurricanes beat working ice breaks, hands down. Once the storm moves on, the weather is near perfect.

We had been warned to be on the lookout for snakes. According to stories coming from the Gulf, snakes, having gained control of the area, were everywhere. They would lie in wait along the roads, within the yards, even inside the houses. We had visions of having to fight our way through a gaggle of serpents any time we stepped outdoors and sometimes even while in.

During the two weeks I spent in Freeport, I saw one snake. He was dead. Snake stories, like fish stories, have a tendency to be exaggerated. Of course once the waters had abated, I suppose those still alive had slithered back into their subterranean hotel or wherever it is snakes spend their leisure hours.

Our boss had a company car, a Plymouth Valiant. In those days it was considered quite small. By today's standards it would probably be classified as medium. One night he suggested that we all go out on the town.

He decided that we would go in his car. Although it must be physically impossible, we piled in, five in the rear seat and four in front. In the back, somehow, four men wedged in side by side. Since I was last in, I lay atop them.

Hitting bottom often, the old Plymouth bounced down the road a couple of miles to a club. No doubt but what Shaky had scouted around during the day.

We went inside and had a beer or two. When the band heard that we were Okies, they played and sang Oklahoma Hills. Eventually we decided to move on. We rode a ways and spotted a club, The Golden Pheasant. Looked interesting. Shaky whipped the car into the parking lot and we moved inside.

The place was filled. Absolutely packed! Except for nine very pale faces from Oklahoma, every person in there was black.

Even Oklahoma blacks were a mystery to most of us. These we certainly didn't understand. Several men eyed us for a moment and then moved en masse in our direction. I figured we were about to be slashed with razors or, at the very least, forcefully evicted.

They were civil. Although I'm sure that in their hearts they did not welcome us, they did tolerate our presence. Eight of us were ready enough to cut our visit short and get the heck out.

Not old Shaky. Nothing to do but what we all sit long enough to drink one beer. They wouldn't let us pay. Nobody sipped. It was time to chug-a-lug. During our tenure, it became deathly quiet. They were probably as suspicious of us as we were of them. Not many white folks came a calling. We could have been the first ever.

We did make it back to the motel. That was the only time all nine of us ever rode together in the Valiant.

Of course everyone had to roll out early the next morning and face a new day. One man, from Oklahoma City, was K. J. Ward. He always seemed to have some witty remark handy that would help us forget that we were tired and longing to go home.

Several of us wandered into the cafe, ready for breakfast. Talk about a quiet, gloomy bunch. I doubt that Bob Hope could have gotten a giggle out of any of us. K. J. did.

There we were, seated at a table, wrapped in a funeral silence, when this pert little waitress bounced over to take our order. How any person could be so cheerful at that time of morning is mystifying. She was all smiles and sunshine.

K. J. lifted his head, stared at her for a moment and in a gruff voice, said, "Burn the toast, pull your hair down over your eyes and growl at me. I'm homesick."

We all busted up. Including the waitress.

As reported earlier our boss was reputed to have been a Navy veteran from World War II. For sure he was a patriot.

Castro had recently conquered Cuba. It was now evident that he was a hard-core Communist as well as a Russian puppet. The mere thought of that Red Rat ruling within a stone's throw of these United States was enough to disturb Shaky, as it did us all.

While we were in Freeport and the rest of us were spending time working or planning an escape from the Golden Pheasant, Shaky's mind, in addition to dwelling on these matters, was turning in a direction of its own. Whether he was serious or just voicing his frustrations, I don't know. Whatever, one morning he approached some of us and whispered his plan. I suppose he wanted to test this new proposal before committing himself to the entire crew. Here is what he hoped to accomplish:

First the nine of us would commandeer, borrow, beg, or steal--anyway, we would take possession of one of the vessels still afloat. With our boss at the helm we would travel to Cuba, find Castro, assassinate the red devil, and then return to the States as conquering heroes.

The plan did have its appeal, but upon close examination it was evident that it also contained a few flaws.

First, it is doubtful that we would have ever found Cuba. If our luck held, we wouldn't. Okay, let us assume we were unlucky. If we survived the landing, it's for sure that we couldn't have travelled far inland and for darn sure we would not have set eyes on Fidel unless he decided to attend our execution. Most likely we would have sunk at sea, been killed on the beach or captured and stood against the wall.

We all know that such events happen every day in the movies. Unfortunately, this is the real world. Still, you have to admire old Shaky and his patriotic spirit. Had his crew been willing to take a little risk, the world might have been spared Castro's insane antics for these many years.

After restoring service in Freeport, we headed home. No convoy or photographers this time. The public relations folks had long ago moved on to more timely topics. Carla was old news.

We made it as far as Dallas and stopped for the night. After checking into a hotel and eating supper with the rest of the guys, I called home to let Dixie know I would be in sometime the next day. We whispered sweet endearments for a few moments and then said goodnight. What happened after we hung up, I didn't hear about until arriving home.

Dixie had answered the phone in our bedroom and sat on the bed while talking, bare-footed, her feet up. After our conversation ended, she swung her feet to the floor and stepped on an ill-tempered centipede.

In a panic, believing the sting of a centipede was deadly, Dixie phoned my parents. My dad called our family doctor who assured him that although the sting was painful, it was by no means fatal.

Soak the foot in a solution of water mixed with salt and soda, rub with alcohol, and all will be well. By the time I arrived home the next day, everything was okay.

Texas recovered from Carla, Castro remained on his throne, and we all returned to work.

EIGHT

A SWEET/BITTER STEP

September, 1963. Tommy Morgan called me into his office where I was introduced to Mr. Ernest Pitts, the Construction Superintendent for the Western Division of Oklahoma. I knew him by reputation only, that he was a hard-nosed man. Mr. Pitts asked questions. Some were personal. Although I wasn't sure of his motive, I know now that he was sizing me up. Then he offered me a job as unlocated construction foreman in his organization.

Pure euphoria filled me to the top and maybe even overflowed just a little. I had arrived! Twelve years after becoming a member of Ma's family, I would soon be one step up that cherished ladder.

Did I jump at the chance? You bet I did.

Since I would be working out-district, where the cost of living was less than in Oklahoma City (A myth perpetuated by the Company to keep from paying folks in small towns as much as they did their city cousins), my basic salary as a supervisor would be only thirty-five dollars more a month than my installer pay. There would be no paid overtime except under extreme conditions.

On September 16, 1963 (Promotions and management title changes were always effective on the first or middle day of the month), I was to report to Lloyd Pierce, District Construction Foreman at Shawnee.

As an installer, I was bareheaded and wearing a short-sleeved shirt during this meeting with Mr. Pitts. He made it clear that construction foreman were required to wear both a hat and long sleeves, buttoned at the wrists. We must set an example for the troops, you know. Being so anxious to jump the fence, I readily agreed.

After all, low-cut shoes were still permissible.

Even though promotion meant that we would eventually have to move from our home, Dixie, for my sake if nothing more, rejoiced at the news.

Earlier that year I had bought a new, Volkswagen Bug. The smooth talking salesman had droned on and on about its construction. According to him, you had to roll a window down slightly before closing a door. If not, there was always the danger of blowing out a glass

"Drive a cork up the tailpipe and she'll float all day," he'd said. Years later when the commercial on television showed a similar Bug plunge into a lake and float merrily across, I would remember his sales pitch with some cynicism. Mine would have sunk in about six seconds. More on that later.

The times, they were a changing. The Company didn't bother to buy me a one-way bus ticket from Guthrie to Shawnee. Ma didn't care in the least that I drove my personal car to my new work site. So far as she was concerned, I could drive the danged thing on Company business until the wheels fell off. Of course if I expected reimbursement or thought that Ma would put the wheels back on, I would be sorely disappointed.

On that fine September morning, I jumped into my Bug and chugged to Shawnee, some sixty-five miles from home. After some searching, I found Mr. Pierce's office, near downtown, and for the first time ever met the man.

Since leaving construction, I had lost contact with the organization. Even when I was a lineman, Slick was the only District Foreman who had made himself visible to the troops on a regular basis. Lloyd carried the same title, but he was quite a contrast to old Slick.

If Mr. Wilkerson was a finesse boxer (and he had the characteristics), Mr. Pierce was a slugger, a body puncher. If Glenn was aloof (and he was), Lloyd didn't mind jumping into the pits and dirtying his hands. Where Slick had made an effort to put me at ease, I sensed soon after meeting Mr. Pierce that he didn't want a young man out of the big city with five years in installation and no managerial experience forced upon him.

I felt as though Lloyd did not like me and had not liked me even before we met, since I was both a new and inexperienced supervisor. Although nobody ever said, I sensed that I wasn't at the top of his list for the job. In fact, there is a probability that I wasn't on his list at all. It was rumored that he had his own man he'd wanted to promote but for reasons unknown to me was overruled.

Lloyd managed the perfunctory introduction, then said I would be his construction foreman in Shawnee. Tom Ely, his cable-splicing foreman, was loaned into Oklahoma City where he was teaching climbing school. Red Guest, his located construction foreman, would handle Tom's job. As for yours truly, I would replace Red while Red replaced Tom and after that God alone knew who would go where.

That same morning I met Lloyd's clerk, Louise Dixson, an affable redhead. It was clear that Louise played a very large role in running this organization.

As it was with most of the older work centers, Shawnee's was located near the railroad tracks. At the time it was built, supplies were shipped by rail. Linemen and cable splicers, installers and repairmen, all reported here. Lloyd informed me that I would be responsible for three two-man construction crews then drove me out and introduced me to Red Guest, the man I would replace.

Old Red was a likable character. Still, upon our introduction, I sensed that he welcomed me with the same enthusiasm as had our boss. His attitude was cold enough to cause frost bite. I couldn't figure it. I had barely arrived. What had I done to offend these men?

Lloyd didn't linger. Having other fish to fry, he instructed his temporary splicing foreman to show his new construction foreman the ropes.

We shared an office. When Ma's children from other departments drifted in, Red introduced me without enthusiasm. By now I was feeling uneasy. Why? The question continued to haunt me throughout the day.

I analyzed the engineering prints. When I would ask about the status of a certain job, Red's reply was, "You're in charge now. Don't be asking me."

Hmmm!

At quitting time the six linemen assigned to me returned to the nest. I stumbled through the procedure of getting acquainted without an introduction.

Something was amiss. Instead of three two-man crews, I had two three-man crews. When I asked why, Red pointed to an old truck that looked as though it belonged in the nearest junkyard and said something about it was due to be replaced in the near future and he didn't want to spend Company money on repairs.

Made sense.

Upon leaving for the motel, I had mixed feeling about having accepted Mr. Pitts' offer. Still, the day had gone reasonably well. It didn't seem as though the job would be all that difficult.

None of my management jobs seemed difficult that first day. There were two reasons for this. First, I didn't know enough about the job to recognize the potential problems. Second, once you've moved to a new job in first-line management and seem to be doing well, old Ma, bless her blessed memory, had a tendency to pile on a little more, and then a little more after that, to see how much you could bear without breaking. And so, wrapped in a blissful ignorance, I felt good about that first day.

Day two. For those folks newly promoted to management, facing strangers who report to you that first morning is nearly as nerve-wracking as climbing your first pole. Sure I'd met them the evening before, but only in passing. In they came, six good men. They didn't seem overly enthused about having a neophyte supervisor. Country folks are a little suspicious of city boys.

City boy! How had that happened?

Somehow or other I stumbled through. The men soon went their merry way. I could go on with what I was supposed to do. What was I supposed to do?

Lloyd Pierce wouldn't tell me; Red wouldn't offer a clue. This made me wonder if the two of them had conspired against me. Yeah, I know, that's paranoia. In reflecting, I'm quite sure that wasn't the case. It seems as though Ma's method of training new supervisors was to throw the poor soul into a den of wolves. Did he cover his head and roll into a ball in a vain attempt to protect himself? Or did he fight back with every ounce of his God-given strength?

To survive, you have to either be as tough as the wolf or try and outsmart him. How I made it, I'll never know. I didn't have enough sense to roll into a ball or I probably would have done so.

There was more to the job than supervising men. Each outdoor supervisor was issued a company check-book on which he could write checks to pay bills so long as they were a hundred dollars or less. He was to pay all bills incurred by himself or his crew up to that amount.

Each bill was forwarded through the District Office with proper account code affixed. Accounting would then separate expenses and keep a running record of expenditures. Our Bible was the Plant Accounting Handbook. Many of the instructions contained within were about as clear as some of those Old Testament books. Even old timers often argued about what code to use for a particular expense.

There was also the Book of Joint Practices, so called because it applied to all Telephone Departments. This covered subjects from who was responsible for what, to overtime pay for supervisors. Not that there was ever much of the latter.

It was soon apparent that my greatest (only?) ally was Lloyd's clerk, Louise. She patiently informed me of my duties, showed me how to write Company checks, what account codes to use and how to humor Mr. Pierce.

Old Red finally came around. As it turned out, he didn't resent me personally. What he hadn't liked was being plucked from his job and being moved to splicing foreman, temporary though it might be. I could see his point. We soon became friends and he even began answering questions.

As for the training of new supervisors, Red, a lay Church of Christ preacher, had this theory: "It's like being converted and receiving the Spirit. When you're given the title, it is assumed by those in the Ivory Tower that knowledge comes in an inspirational flash."

Mr. Pierce kept his distance, evidently seeing if I would survive on my own. One day, however, he arrived at the work center right before quitting time and saw the two three-man crews drive in. He asked rather forcibly why I was working three men to the crew.

I told him the third truck needed major repairs. His reply, as I understood it, was, "Get it fixed! I want you back to two-man crews as soon as possible."

Next morning I phoned the Ford Garage and had them tow the old wreck to their repair shop where with my approval and consent, they put in a rebuilt engine. When I later received the bill for six hundred and some dollars, without giving the matter serious thought, I attached the proper voucher and forwarded it to the District Office. When that bill hit Mr. Pierce's desk, all hell broke loose.

Yes, it did.

By phone he ordered me to his office. Right now! From his tone of voice, I was certain it wasn't to discuss a raise in pay. Within ten minutes we were face to face. He was livid. He grasped the bill in his chubby fingers, so angry he had difficulty even speaking.

The Company monitored what it spent on motor vehicle repairs. The authority for a first-line supervisor was limited to the ridiculous sum of only ten dollars; a second-line such as Mr. Pierce, could authorize no more than a hundred. Even Mr. Pitts, in all his glory, reached his limit at five hundred. Lloyd demanded an accounting.

"You told me to get the truck fixed," I said.

"I told you to get an estimate!" he shouted.

"You told me to get it fixed!" I shouted back. This exchange continued until I stomped out of his office, not knowing what would happen next and not really caring. One thing seems certain though: had I responded in a timid or obsequious manner, he would have been on me unmercifully from that day on until the eventual end of our association. The kicker is: once the dust settled, our relationship actually improved. These guys explore for any weakness and attack.

As the weeks passed, the job pressures built. Each of the District Construction Foremen fought to be top dog. For them to be, their supervisors and crews had to be. To reach the top, your indexes had to be higher than anybody's. It wasn't until after I'd spent some time in management that I became acquainted with the importance of paper performances.

What work you or your crew did wasn't nearly as important as what your indexes said you did. Some first-line supervisors became pencil pushing wizards.

Safety was a number one priority, as of course it should be. A close second was an index known as man-hour performance or units per hour worked. For some but not all work operations, labor units were credited to the construction foreman.

During Mr. Pierce's supervisor's meetings, those with high man-hour performances received a goodly number of attaboys. Those who came in below par, for whatever reason, became victims of his wrath. He was a man who wanted results, not excuses. Even though you might consider your reasons valid, this carried little weight with the old man.

On that infamous November day when John F. Kennedy was assassinated, Lloyd was having his monthly supervisor's meeting. While eating lunch at a restaurant, we heard the news. Of course we were all saddened and apprehensive, but Lloyd was devastated. We spent the remainder of the day listening for further details.

In December, Louise informed me that I was to purchase the annual tags for my trucks, company car, and, while I was at it, would I get Lloyd's? Sure. I picked up the titles and bought tags the same day.

Lloyd was not happy with his new tag number. Had murder not been a capital offense, he would have surely strangled me. I was probably the only person in the Shawnee District who did not know that the license tag number on his company car was PT-109.

I didn't know to ask for the number. This hurt the old man more than had the high cost of truck repair. That was only money. Here he had lost the PT boat number of his hero, John F. Kennedy.

The first of January, 1964, I was transferred to Seminole as construction foreman, still unlocated. This move put me even farther from home.

In those days our buried cables and wires were placed by contractors. These jobs were bid on by a list of folks who met the company's requirements. Low bid did the work.

Unfortunately, emphasis was often on quantity instead of quality. The lower the price, the higher your index. The higher the index, the more goodies came your way. Consequently, some of our cables weren't buried too deep.

If you insisted on a proper job, the cost would be higher on subsequent orders. The Maintenance Department, responsible for cable once it was put in use, often seemed to be at war with Construction. The two had conflicting priorities.

One weekend I stayed Friday night at Seminole to meet with a cable-laying contractor on Saturday. Once my business was finished, I headed for home in my little Volkswagen. Before reaching Shawnee the rains came.

When I drove into town, water was running deep. The rain stopped but the streets were still flooded. Feeling confident, the salesman's words still fresh in mind, I drove my Bug into the flood.

Water rose over the tailpipe. Ahead of me a car stopped and I had to do the same. The engine died and of course wouldn't start again since the exhaust system was submerged. Ever rising water came up on the doors and began leaking in. Soon the deep floor well was full. I thought bad thoughts towards a certain Volkswagen salesman. Had he been there, I would have tried to hold his head under until he quit bubbling.

A pickup pushed me out of the deep water. I got the engine started and sloshed home where I spent the next several hours drying out my nearly new car.

By spring I was feeling pretty comfortable with my job. Other than for being away from home during the week, it was bearable. Finally things were going my way.

And then Comer called.

NINE

CHIEF OF THE WIRES

A Kansas City undertaker is reported to have invented the automatic dialing system in the 1890's while looking for a way to bypass telephone operators who he suspected were routing his calls to competitors. By the time Comer called most local calls were being completed by this method.

Comer Williams was the chief over all telephone service for the Shawnee District. He was unique. Most certainly he was one of a kind. Comer differed from any man I've ever met.

Comer was absolute lord over his kingdom. His throne was in Shawnee, and his reign covered all towns from Chandler to Holdenville and from Guthrie to Ada. He was a third-level supervisor, responsible for all of the indoor and outdoor telephone facilities that fell within his domain, other than for those delegated to Mr. Pierce.

Their offices were on the ground floor of the same building. Their clerks, along with Mr. William's secretary (she too held a management position) shared a large room. The relationship between these two men was about as amicable as that between two roosters in a pit.

Since they were separated by lines of organization, the fact that Comer outranked him didn't cut any ice with Mr. Pierce. Lloyd lived in a world of his own, marched to his own drummer and Comer's indexes be damned!

From Comer's viewpoint, Mr. Pierce turned his head when cable contractors deliberately cut corners and plowed the cheapest route, thereby enhancing indexes that were precious to the Construction Department. Mr. Pierce defended his position with all the tenacity of a she-bear with cubs. Sometimes the fur flew.

One day in May when I answered my office phone Comer's booming voice rolled out of the receiver. I first thought he couldn't find Mr. Pierce and was after me for some real or imaginary transgression committed by the Construction Department. In other words, I immediately went on the defensive.

After a few pleasantries, uncharacteristic of him, he said, "How would you like to be my Wewoka Wire Chief?"

"Be what?"

"You'll get a thirty-five dollar a month raise."

"Will I still be paid for room and board?"

I'd hit a nerve. "Hell, no!"

"It sounds as though I'll be taking a cut in pay then," I said. "I'll think it over and let you know in a day or two."

"There's not a damned thing to think over," he said. "Your title was changed this morning."

"That being so, why did you ask me?"

"I thought you'd say yes. Be there Monday."

"Slam!" went his handset. He was a busy man. No doubt but what those seconds he'd spent enhancing his 'good old boy' image had cost him dearly.

I confirmed Comer's invitation with Mr. Pierce and had no choice other than to report to Wewoka on Monday morning.

As Wire Chief, I was responsible for the central switching offices and all the outside wires and cables and telephones in six towns. Besides Wewoka and Cromwell, a little town nearby, I was in charge of Chandler, Luther, Wellston, and Carney.

Soon after reporting to my new job, we cut the central office at Chandler from step-by-step to crossbar. We had an open house for the public. One man who assisted me was Bill Fine. When the people came through, he would say, "My name's Bill Fine. I'm fine, how are you?" Clever, huh?

A few years later, Bill was promoted, soon made second-line and for a while, when I was an installation foreman, I was reporting to him.

Three men reporting at Chandler had to do all the telephone work in four towns. I was expected to spend time there, supervising, sixty miles or so from Wewoka. And yet I had no company car. Hmm!

When I questioned Comer as to how I was to travel from here to there, I received a vague response.

The job at Wewoka was sometimes used as a training ground for new wire chiefs. The man I replaced was a college grad who'd been hired into management. He, like others of his ilk, knew that more promotions lay ahead and so he had driven his own car, a Thunderbird, without expecting or asking for compensation.

Eventually he moved several promotional steps up the managerial ladder, so I suppose this expensive habit paid dividends. We had four kids in school and didn't feel as though we could afford to donate our personal car to the Company. It wasn't as though Ma was destitute and couldn't pay her own way.

My office was near downtown, located in the same building as the switching equipment. First morning on the job, I stepped out the back door and in came Comer. Yes, he had a company car. A full sized Chevrolet. He parked, crawled out, and greeted me with words loud enough to frighten birds for a six block area.

"This place is a mess!" he roared. "Why don't you keep it clean?"

For a moment I was too flabbergasted to answer. I know now that even as Lloyd Pierce had tested me early on, Comer also was interested in how I would react.

Well, surely the man would listen to reason, so I tried to explain that I'd hardly had time to find my office, much less hire someone to clean the parking lot.

Not good enough. He roared again. Only one thing to do then and so I did it. Roared back, that is. This seemed to satisfy the old devil and made my tenure with him more bearable than it would have otherwise been.

He informed me that the company would pay my meals and lodging until I moved my family. He suggested (rather forcefully) that this move take place within the next two weeks. Wire chiefs were always programmed to be active in community affairs.

I rented a house in Wewoka and we rented our house in the country to another family. It was good to be together again. Dixie enrolled the kids in school. Ted, so recently our little baby, put a basket on his bicycle and started throwing the daily paper. This lasted until he wrecked his bike during an ice storm. He was ready to keep on keeping on, but mamma wouldn't hear to it.

Such is the way of mothers.

The company had paid my moving expenses, we were once more together as a family and all was well.

Or was it?

The problem still existed of how to get from one place to some other place in order to effectively supervise. Since we needed a larger family car, the Bug was traded. When I needed to take care of business, whether in Wewoka and Cromwell or Chandler and its satellite towns, I drove my personal car, kept an accounting of the mileage and, after a few weeks, wrote myself a check on the Company's account. This was in compliance with one of the Company's joint practices.

I attached the check carbon to a voucher, along with a full explanation of what, why, and how come. I forwarded this bundle to the District Office for belated approval. When it reached Comer's desk, his reaction was similar to what Mr. Pierce's had been upon receiving the unauthorized bill for motor vehicle repair.

He told me that I could not, would not, under any circumstances be compensated for driving my personal car on Company business without approval from his boss, the Division Plant Superintendent, and he, Comer, wasn't about to request said permission because the answer would be no, and that was that.

"How then do you expect me to conduct the business of the Company without a car to drive?" I asked.

"Use your imagination. Just don't send me any more vouchers for the use of your personal car."

Comer hung up before I could formulate a suitable reply. Words always come too late.

Within my thick head, the rusty cogs were soon meshing. I made a silent vow to never again drive my personal car without compensation. It was obvious that I would not be paid one more cent if I did. Not by Comer. Having been officially warned, a repeat offense would be insubordination.

My next stop was Wewoka's only cab company where I arranged for monthly billing. So, when I needed to do business in Wewoka, I had a chauffeur. If I needed to go to Chandler, I would ride a cab the ten miles to Seminole.

Gene Flannigan, a likable cuss with whom I'd become acquainted while working as a construction foreman, was Seminole's wire chief. I would dismiss the cab, borrow Gene's company car and head for Chandler.

My cab fare was considerably more than would have been the payment for the use of my own car. Comer approved the bills without complaint. Some months later, a new car did arrive for his Wewoka wire chief. By the time it arrived, I was long gone.

Our organization was indexed on so many performances it was necessary to compile a log to be certain none were missed. Being the only supervisor in town, I was responsible for all that happened. And for that which should have happened, but didn't. Yeah, we had sins of omission as well as sins of commission.

Each month, after receiving the myriad of index reports for his district and for individual supervisors, Comer would call a meeting. He would accept no excuse for being absent. Comer would rant and rave, swear and shout. I could not believe that a man of his position would conduct himself in such a manner. Initially it was upsetting; later it was mostly amusing.

Comer covered every man's indexes and little time was wasted dishing out attaboys. If your indexes were the best in the entire world, that is what you were being paid to do. If not, why not? If his indexes happened be at or near the bottom, a supervisor would be berated in plain view of his colleagues.

At Wewoka, I had three outdoor craftsmen and one who handled indoor work. Those outdoors were responsible for installing and removing telephones and for maintenance of all wire and cable. They were stretched beyond any reasonable limit, as were the three men who handled Chandler.

Jerry Suva was my indoor man at Wewoka where we had an antiquated step-by-step switching office. I all but left Jerry alone. What I knew about central offices could have been written on the head of a pin with room enough left for the Lord's Prayer.

The same was true with cable repair. A lesson you learn early on is to depend on your skilled people to carry the load. I depended; they carried.

When you're new in management, you "wing it", to use an old expression. You move ahead by trial and error, making every attempt not to repeat mistakes. You tend to believe everyone is against you. Not so, McGee.

Even Comer, when you peeled away his facade and looked beyond his contrariness, was human. He cared. For the company he cared deeply, and I do believe he cared for his subordinates. However, any man who rolled himself into a ball to prevent being injured would be continually kicked by Comer. Steel is tempered by fire and in his opinion that is what he was doing.

Whatever folks in the telephone business called him, and much of it coming from his subordinates was less than complimentary, he was a devout Company man, totally dedicated. The number of hours spent working in Ma's interest meant nothing to him.

The Bell System Practices (known to the user as BSP'S) are a collection of books comparable in size to the Encyclopedia Brittannica. They were standard issue by AT&T so that a specification in Oklahoma was no different from one in New Jersey.

There were some local practices but not enough to confuse the initiated. Comer carried a set of BSP'S in the trunk of his car. They were as important to him as the Holy Bible is to a born-again Christian.

If he said a job was to be done a certain way, you had better believe it. To argue was futile. Many tried; few succeeded. He would quote chapter and verse, whip open the book and show you the error of your ways. With few exceptions, Comer was correct.

His words of wisdom to me were: "Never try to cover up. If you make a mistake, admit making it but don't repeat it."

Being a wire chief exposed me to all types of new situations as well as to various and sundry people. I was expected to join civic clubs; I did. I was also expected to be a stalwart citizen; I probably wasn't. It seems as though I was too busy learning my new job to pay much attention to my civic duties.

When we moved to Wewoka I had a bell chime installed along with several telephones. Chimes were new and they were considered to be neat. Of course they were off-key and not too melodious. Call comes in and it goes, "Ding-dong!" In the beginning I had to proudly show off this new addition to our house. Soon I hated the darn thing.

For the first time ever I was on call twenty-four hours per day, seven days a week, all the weeks of the year. Even vacation time was not exempt. Just like a doctor. Only thing is, his calls are intercepted by an answering service.

The Chief Operator was located at Holdenville. If I wanted to take the family away for a little recreation, she needed a number where I could be reached. Going fishing, going to a movie--tell her.

The reason she was interested in my travels, as well as that of all other wire chiefs, was that alarms for Wewoka and Cromwell terminated in her office. Our switching offices were unmanned nights and weekends.

Whether we were asleep or awake, telephone cables were outdoors, exposed to the weather. If a cable got wet, an alarm sounded in Holdenville. If the switching equipment skipped a beat, ditto. We had a buried toll cable under air pressure that passed through Wewoka and terminated in our office. Let the pressure drop on this little beauty and folks everywhere got excited.

When an alarm sounded for whatever reason, at high noon on a Saturday or at two o'clock any morning, the chief operator on duty would pick up her phone and dial my number. One in the morning: "Ding-dong!" Just as I'm falling asleep: "Ding-dong!" While watching football on a Sunday afternoon: "Ding-dong!" Many the time I wanted to rip the danged thing off the wall and toss it into the trash.

After taking the report, I had to call a man out to fix the trouble. If the problem was serious enough, I was expected to crawl out of bed and do a little supervising.

One of the first edicts Comer laid down to this inexperienced wire chief was: "If you have cable trouble any time of the day or night, I want to know about it. Call me at home if need be."

The first time it happened, I did just that.

A summer thunderstorm hit Wewoka late one night while I was sound asleep. "Ding-dong!" An alarm had come in.

I called Jerry Suva who went to the central office, identified the trouble and, "Ding-dong!" "The cable's wet going out to the radio station," he said.

I crawled out of bed, phoned Ralph Miner, who was an experienced outdoor repairman, and headed for the office. I dialed Comer's number and disturbed his dreams. Whether or not his home telephone went "Ding-dong!" I don't know.

However, it was soon clear that the old man was not happy to learn of my problem. His reply was, "Have the damned thing fixed in an hour and call me back!"

I crawled into the truck with Ralph and away we went. The prescribed method for finding wet cables is to hook the leads from a special set, one on half of a wet cable pair, the other to ground and shoot voltage on the line. According to the book, the wire will then burn to ground. Then you hook up a tone and follow the sound until it fades away. Bingo! You split open the cable sheath, dry out the wires and make your repairs. Then you call Comer after which you go home and crawl into bed.

This one was particularly troublesome. Two hours passed and we hadn't located the trouble. We returned to the office for something or other and Suva informed me that Comer was phoning every five minutes, demanding to know why we were taking so long. He'd insisted that I call at the first opportunity.

When Comer answered, he wasn't in the best of moods. One of his (and mine) indexes was time taken to clear cable trouble. We exchanged words and then back out with Ralph I went. Another two hours passed before we found and cleared the trouble.

Weary, we returned to the office. I dialed Comer. He did not say, "Well done!" Never again did I phone him when we had cable trouble. We would report it to the District Office on our morning report and that would be that.

All good things (and some not so good) must come to an end. For personal reasons, after nearly a year in Wewoka, I needed to return to Oklahoma City. The only vacancy was for a construction foreman.

Oh well, I'd done it before. At least I didn't have to listen to any more "Ding-dongs!" during the night.

TEN

CHASING THE PAST

In Oklahoma City, my immediate supervisor was an old-timer whose office was located several miles from mine. If I'd had my way, we would have been separated by at least another hundred miles. He surely felt the same way. Some folks seem to have a natural conflict. We did.

His clerk, Lucille Daggs, in true tradition, was the dynamic force that kept the wheels turning and running smoothly. A department could survive with an inept first or second-line supervisor but take away a clerk who's both knowledgeable and capable, as was Lucille, and everyone involved is in deep trouble.

Three other supervisors were working out of my new reporting center, none of whom I knew other than by hearsay and reputation. Those workmen I had always considered to be prima donnas, PBX installers, reported to V. B. Sweatt. Shoot, they were just ordinary folks. Not that much different from any of us.

Mr. Sweatt, several years my senior and with much more experience, was a good old boy. He was one of those guys who looked relaxed but ran a tight ship. Every move he made seemed to pay dividends.

Now and then I would meet a man blessed with extraordinary ability and yet, for whatever reason, had never advanced beyond the level of first-line. Mr. Sweatt certainly fell into this category. Only thing I can figure is he was too plain spoken. He always spoke his mind and didn't kowtow to the big cheeses, whoever they might be. Also, he refused to participate in the politics that are inevitable in any large company.

Sweatt considered time away from the job to be his and did not join the Company Bowling League where he could buy beer for his superiors, nor did he take them hunting or fishing. These activities often seemed to be prerequisites for promotions. Some first-line supervisors seemed to spend more time greasing the wheels than they did in running their jobs.

Too often this method seemed to pay dividends. Sweatt did not want a promotion bad enough to pay that price. Chalk up another admirable quality for V. B.

I don't mean to imply that everyone promoted beyond first-line management was a boot-licker or that their nose had a brown tint. Sure, it was obvious that many who succeeded fell into this category. But the installation foreman was Bob Reynolds and the way I saw it there wasn't a kowtowing bone in his entire body.

Bob did receive a well deserved promotion to the second-line level at a later date. At the time we worked together, he was not only responsible for residence and small business installation, but also for maintaining the large work center where we reported.

He handled repairs to the building, paid the utility bills, and called the exterminator when needed. Bob was also responsible for the house-service contractor who sent folks in after hours to do the cleaning.

The Company had its own buried gasoline tanks so each weekday evening a man contracted to do so would fill the myriad of trucks needing gas. Guess who was responsible for seeing to it that the storage tank didn't run dry, for dipping said tank before and after filling to be certain we weren't shortchanged, and for compiling the gasoline tickets monthly on each vehicle to see who was burning all the petrol. You're right: Bob was.

Besides, he was responsible for the storeroom (other than for construction; I was responsible for my own) and seeing to it that they stocked the multitude of equipment needed to install and repair not only residence but large business systems. All telephones coming into or out of the storeroom were counted and they had better balance or come darn close at the end of the month.

Bob, as were all outdoor supervisors, was even responsible for the maintenance and upkeep of those motor vehicles assigned to his crew. Each supervisor was indexed on motor vehicle expense.

The company policy was to drive a truck for ten years or a hundred thousand miles, whichever came first. For city trucks, ten years was usually the criteria. Short runs, along with stop and go miles, take their toll. If money was tight, another year or two might be coaxed out of the old wrecks. Money was nearly always tight. We were constantly coaxing.

Bob was responsible for so many duties, I often wondered how he found the time to supervise. Ma believed in getting her money's worth out of her children.

R. O. Wall, a prince of a fellow, was the repair foreman. A large man, R. O. could come on strong. He could seem overbearing and obstinate as he tore into a subordinate, ranting and raving.

The craftsman, though, recognizing this as smoke and knowing that he was dealing with a boss who had a Teddy-bear complex, a man who would have literally given the shirt off his back, would rail back, exchanging tit for tat. It was mostly for show.

R. O. cherished every moment of these raucous exchanges.

Following a verbal joust with Mr. Wall, many a man would drive out of the work center with a smile on his face. Most of those who reported to him, would have gone to war for their boss.

Mr. Wall loved to eat and it showed. There were occasions when the four of us would be in our offices at noon. If so, we would usually go to lunch. Since I had the best Company car, a nearly new Mercury Comet, I was elected to drive. Indeed, other than for an old wreck driven by R. O., mine was the only Company car in the work center.

Wall loved Mexican food and fried chicken.

Near us was a Mexican Restaurant, El Charito, a popular establishment with a large selection of food from which to choose. Occasionally we would go there.

A bowl of hot sauce would be sitting on the table. R. O. would pour its entire contents over his heaping meal and to the waitress say, "Honey, you'd better bring these other boys some of this."

He'd take a bite, a smile would spread his face, and he would say loud enough for half the folks in the restaurant to hear, "Whoee! Wake the kids, mamma, and bring them to the table. I ain't never in my life tasted nothing as good as this."

How he was able to eat such hot food, I don't know. Not only did he eat, but he relished every bite.

One summer day, following such a meal, some fifteen minutes after returning to the work center, I went outdoors and there was R. O. He had a scoop out of the ice machine. Water had been mixed with ice and he was pouring it down.

"Gotta put out the fire," he gasped, between huge gulps. This didn't daunt his desire for hot food.

His favorite place to eat fried chicken was a little cafe at Jones, Oklahoma, a small town east of Oklahoma City. Their telephone exchange was owned by another company. This put us slightly out of our district. Even so, we'd occasionally slip over there to eat.

This little cafe was antiquated, a "ma and pa" type establishment. Ma worked the kitchen while pa, dressed in overalls, served the tables. They fixed the best fried chicken I've eaten other than for my wife's and mamma's.

We would each eat a heaping helping and R. O. would order a sack of chicken to go. This he would eat on the way back to the work center. Sitting in the back seat of the Comet, he would devour all but the bones which would soon be flying out the window, making many a scrounging dog happy.

R. O. had a business on the side. This practice was not exactly encouraged by the Company but so long as you didn't conduct your business on their time they seldom ordered you to choose between it or them. In the case of a first-line supervisor, Company time was usually considered to be twenty-four hours a day, seven days a week. At the very least, it was difficult to determine when their time stopped and yours started.

Wall bought and sold large, used tires.

He drove a red truck with sideboards. Many days he would park out front and his truck would be loaded with old tires. After work that evening he would head for his tire yard, near Nicoma Park.

Although he was deep into the business, so far as I could tell it didn't interfere with his job. Oh, he might make or take a tire-call in his office now and then, but so what? Being a repair supervisor, he took plenty of Company calls while at home on nights and weekends.

One morning after the crews had been sent on their way, Sweatt and Reynolds decided to play a little trick on their buddy. Reynolds slipped into a room down the hall and dialed Wall's number. R. O. was talking on another line at the time and so Sweatt answered at his desk and said, "Tires? Tires? Why are you calling? We don't deal in tires here. This is the Telephone Company."

And then he slammed down the handset and said he didn't know why people would be calling the Telephone Company for tires.

Needless to say, old Wall came out of his chair, squealing like a punctured panther.

By now Dixie'd had enough of staying home with four kids who were sometimes out of control. She went to work as a copywriter at John A. Brown's. Since I didn't need our car during the day, she would drop me off in the morning and pick me up in the evening.

One afternoon the temperature dropped fast and freezing rain covered the roads. They let Dixie off work early; she picked me up before five. I drove us onto the Broadway Extension. Traffic was at a standstill.

Tires spun. Cars slid off the road. About seven, we rescued a young couple whose car had run out of gas. Along with the other man, I was usually outside the car, either pushing others out of the way or trying to keep our car moving.

Ten o'clock that night. In five hours, we had travelled some eight miles. South of Edmond, all traffic came to a standstill. My gas gauge was on empty. There was nothing to do but lock up for the night. Fortunately, we were near a large motel.

The couple we had picked up on the road took the next to last room available. We took the last. I called home and told the kids what was happening. By now they were teen-agers and capable of staying alone.

We hoped.

The next morning we went into the motel lobby and were astonished. Blankets and pillows were on the floor and people were bedded down in every available space. The day dawned clear and cold. I spent the morning getting our car moving. Neither Dixie nor I made it to work. Heck, it took most of the day just to get home.

So, the story started making the rounds (with my help, of course) that this woman and I were riding together and when the storm hit we stopped at a motel and took the last room available. Not only the last room but the last bed, also. Well, we did.

There were instances when an outsider would have thought that folks in these separate departments worked for different companies. Many were obsessed with building their indexes to the highest level possible. They suffered from tunnel vision. Take care of your own business while others sink or swim. It is no affair of yours.

I didn't see it that way. Nor did any of the three maintenance supervisors with whom I worked at the time. We always helped one another.

Being a construction foreman in the big city was more than a little different from wearing the same title in the out-district. There, working many small towns and the surrounding territory, we had to cover a wide area. Here, we worked in a relatively small space.

At this late date, open-wire lines were nearly non-existent. Two of our primary work functions were pulling in underground cable and changing out poles hidden behind houses.

It was a pleasure to have Wayne Fitzgerald on one of my two-man crews. He was an old hand, a lineman because he wanted to be, not because he wasn't qualified for other jobs.

Unlike most of us who as linemen were anxious to bid on another job once we had seniority, Wayne was content to be where he was and do what he was doing. When a new man would come to the crew, I would team him with Fitzgerald. There the neophyte would receive proper training, a task requiring more patience than most folks possess. You could depend on Wayne.

One afternoon Wayne came into my office and told me of a troubling experience. "I don't think anything will come of it, Don," he said, "but just in case, I wanted you to know."

He and his buddy had been working in the outskirts of Oklahoma City and once their job was done had turned their truck around in a driveway. An inquisitive kitten had come out to investigate, and unfortunately had paid with its life when smashed by a truck tire. Wayne hurried to the house, rang the doorbell and told the lady of the regrettable incident.

Her reply was, "Don't worry about it. We were wanting to get rid of the kitten anyway."

The story should have ended right then and there, but it didn't. When her husband came home and heard what had happened, he decided the deceased cat was an animal with some value and he should be compensated. When a large corporation is involved, there are folks who have a tendency to jump in for all they can get.

Next day he phoned the Business Office and they routed the call to the construction department. The end result was I took the man a company check for thirty-five dollars in payment for a feline he didn't want in the first place. It's a wonder the old boy didn't ask for a thousand.

My boss, the Old Timer, was a stout man, even in his advanced years. Balding and a smoker of cigars, he had been a District Construction Foreman longer than I had known him. To say he was from the old school, is comparable to saying Bill Clinton is a politician.

My only past contact with Old Timer had come when I was a telephone installer. At our work center was a meeting room where the guys gathered before starting time to sip coffee and munch donuts. The floor was a waxed tile. A fellow strolled through early one morning, stepped on a bite of donut someone had dropped, and out his feet flew. Down on his back, and just that quickly the company had a lost-time accident.

All such accidents required that a committee be convened, responsibility assigned, and a determination made as to how future mishaps of a similar nature might be prevented. Old Timer headed up the committee; I was one of the members.

From the Company's viewpoint, all accidents are preventable. If it was a motor vehicle accident, perhaps the man should have been driving on another street. Sure enough, that was often the conclusion reached by the investigating committee.

If a man falls from a pole, it was usually due to carelessness on his part. Either he was climbing on the wrong side, didn't have his gaffs sharp, or something of that nature. If a man slips on a crumb from a donut and falls on his back, hmmm. Well, now.

Without exception, unless it was an act that could be blamed on God, the blame for an accident fell on the victim, his supervisor, or, as happened in most cases, both. This accident was unique. Not many folks slip on a crumb and have to be hospitalized with an injured back.

We convened in a conference room at the work center where Old Timer searched for a safety practice that would cover this particular accident. The result was somewhat vague.

We visited the man in the hospital and listened to his excuses. Yeah, a likely story! Why didn't you stay downstairs where you belonged?

We talked with the man's supervisor, feeling that perhaps he should have made the rounds earlier, policing the area for donut crumbs. Surely doing so would fit his job description somewhere.

It was our job to reach a conclusion and write a comprehensive report. In this report we had to make a few recommendations, no matter how silly or ambiguous they might be. Our conclusion was: Folks eating donuts should pay attention to their crumbs, and those walking in the area should be on the lookout for foreign objects on the floor, no matter how tiny they might be.

So, take that! The poor man's in a hospital bed, in agony, and we burden him with most of the blame.

The next I was to hear of this Old Timer, I was reporting to the old codger. He tried to use that old man, filled with wisdom, approach. You know, "Son,

I've been everywhere, seen it all, so just listen to me and you'll be all right." I don't know why, but most of the time I couldn't take him serious. The way I saw it, he was about twenty years behind the times.

My years as a telephone installer and tenure as a wire chief, had plumb gotten me out of the habit of wearing a hat and shirts with long sleeves. Yeah, I know, I promised Mr. Pitts I would but times change. That is the only way I can rationalize my attitude at the time.

The company, due to cries of anguish from its work force, had relaxed its mandate so that workmen only had to cover hands, arms and head when they were walking wood. Man working the pole had to wear the proper attire; man working on the ground could go bare-headed and wear short sleeves.

The company didn't exactly care for this policy but they were forced to acquiesce.

This went against Old Timer's grain. In his way of thinking, a supervisor should set an example. "Wear long-sleeved shirts every day and every time you leave your office, don your hat, Don."

Tradition!!!

My argument was, "The men don't have to dress this way unless they're swinging in the air. If I skin a pole, I will be properly covered."

He puffs on his cigar, wrinkles his forehead.

"Why in the devil should I wear something that I'm not comfortable in if the company changes its rules for craftsmen?" I argue. "Do you really believe that just because I wear all that garb the linemen are going to want to impress me to the extent that they will do the same? Bull! What they will say is, 'If old Don doesn't have any more sense than that, he's losing it.'"

Needless to say, this attitude hardly endeared me to the old man. He spent hours trying to convince this stubborn farm boy that he was right and I was wrong. Looking back, it's a wonder he didn't have me up for insubordination. Had he been convinced that he could have done so and made it stick, I believe he would have.

By now I was becoming more rebellious than was good for my career. To get along, you gotta go along. This is hard to do, but there are times when even if you think it's the wrong path, it should be done. Tradition dictated that we stick with the status quo.

In spite of our disagreements, the old man liked to sit in my office, or ride with me while presurveying jobs, and tell stories of old. He dated back to when linemen stayed on the road much of the time, moving in large groups from town to town. When doing so, they would room in a boarding house.

He told how they would sit at a long table and how the platter of biscuits would always be placed near his right hand. Someone down the table would say one word: "Biscuit." He would snatch one off the platter and flip it to the hungry man without so much as looking up. I've heard that story a few times.

Yeah, Old-Timer had fought the wars. Perhaps he had been in combat a little too long.

The way he saw it, nearly every person who had started with the Telephone Company during the past twenty years, especially those who worked in other departments, were lazy, good-for-nothings, doing no more than looking for a free ride, a free lunch, or anything else free they could get their hands on.

When he measured others against himself, we all came up short, dreadfully short. He became obsessed with the idea that it was his responsibility to point out the error of our ways and make superb telephone people out of us simple souls, even in spite of our many shortcomings.

As he tried, it must have been comparable to making bricks without straw. He even got down on some of our contractors.

Most of our contractors were decent folks, just trying to make a living. Some did; others went broke. Competitive bidding kept them from gouging the Company. A few might make a decent wage off of us but they didn't get rich.

A story got out about one contractor who worked for us. In our world today, when compared to our young President, some of our U. S. Senators and a few of the TV evangelists, what he did wasn't all that dramatic. Still, he had the reputation for being a womanizer.

Not being a public figure, his episodes didn't make the news. Other than for the grapevine, that is. That line stayed hot.

One day, as the story goes, he told his wife he would be out of town until late, driving a job before placing a bid. That night a conscientious citizen found the poor man's car parked down by the river, seemingly abandoned.

Believing the car stolen, said citizen hurried to the phone and called the sheriff's office. In an effort to solve the mystery, a deputy called the owner's home to see if their car had indeed been snatched. The man's wife took the call, added one and one together and came up with two doing a little hanky-panky. She didn't care for this sum total. So she drove to the river, doused her lights and waited.

Sure enough, late that night, in drove another car. Two people sat there, kissing intimate farewells, and then the man got out.

When the wife flipped on her car's headlights, she caught sight of a woman driving away and of a startled husband looking for a hole in which to hide. Alas, there was no escape. I suppose the temperature inside their home was somewhat chilly for a while.

Finally Old Timer decided that in spite of his best efforts, he could not save the company. He retired, a move that was probably best for us all and him, too.

My new boss was Bill Wells. Relatively young when compared to the man he had replaced, Bill was a ball of fire. He spent the first day or two cleaning out his desk and files, throwing away everything that wasn't essential. It seems as though there had been a lot of paper kept here and there that wasn't essential. Efficiency was his byword.

When you entered his office, you would never see more than one piece of paper on his desk, and that only if he was working on it at the time. Otherwise, the top was clean. An admirable trait, but one that I was never able to master.

Bill and I hit it off just fine. Soon I was to leave the construction department for good.

First though: even more storms.

ELEVEN

THE LAST STORM STORY

When the big blizzard hit New York City in March of 1888, telephone service wasn't restored until May of the same year. The marvels of modern technology, along with improved methods of transportation, have allowed us to lower our clearing time considerably.

I promise, I'll hardly mention storms the rest of the way. Oh, maybe just a teeny bit now and then, but nothing major.

These accounts are not in chronological order but in order of their magnitude. They did occur within a two year period, when I was a construction foreman, back in the second half of the sixties.

First, there was a baby storm that dipped down near Cashion, a little twister that didn't amount to much as tornadoes go. Of course it did manage to knock down a few barns and such, as well as clip old Ma's lines here and there. Late in the afternoon I received a call from Old Timer, telling me to load 'em up and head 'em out.

Since the damaged area was practically in our back yard, we arrived while we still had several hours of daylight. A contractor was unloading new poles, so we dropped in at the first break and went to work. I drove the area to get a feel for what we had to do and then returned to my crews.

An occasional repairman wandered by, looking for broken drop wires or such. Otherwise, it was our baby, our responsibility, to replace broken poles, splice the severed wires together and all that.

By late evening it was apparent that we would be working all night, and well into the next day. I was considering running into Edmond for food when here came Old Timer driving up in his personal car, his nice wife beside him. We chatted for a few moments. Mostly I filled him in on the extent of the damage and how long I thought it would take to restore all service.

When the old gentleman learned of my intent to feed the troops (No more working all night without any nourishment, as we had with Smiley), he volunteered to run the errand.

Sounded reasonable.

At the time there was an establishment in Edmond that featured nineteen cent burgers. Even though this was late in the sixties, nineteen cents was still a cut-rate price. You can imagine what that piddling sum bought.

If you'll recall the Wendy's commercial some years back, where the dear old lady asks, "Where's the beef?" you'll know what Old Timer brought back to sustain working men throughout the night. Onion-burgers, is what they were. If you prefer small buns smeared with mustard and filled with thick slices of onion and enough meat (beef?) to satisfy one hungry mouse, you'd have truly loved these.

I don't know how the old man was with money from his own pocket, but he was right stingy when it came to spending Company funds. Perhaps he paid for these himself instead of using Company money, I don't know. Anyway, he delivered these baby-burgers proudly, as though they were ribeye steaks with all the trimmings, wished us a good night and headed home. The starving linemen looked at supper and muttered mutinous threats. I admonished them to keep the faith, jumped into my company car, and headed for Edmond.

Before dark I returned to the workmen with real hamburgers and French fries and Cokes and all that good stuff. Sometime during the night it's quite possible that a scavenging opossum discovered a pile of onion-burgers in a roadside ditch and done thought he'd found 'possum heaven.

Soon after daylight we drove to a Cashion Cafe and had breakfast. Later that day, when all service had been restored, we headed for home and a long sleep.

Tornadoes come in all shapes and sizes. The one that roared through Topeka, Kansas was a pappy size, maybe a grandpappy. It swooped down at the southwest edge of the city and carved a wide path of destruction as it cut a diagonal path right across town. The debris had hardly settled before Ma was organizing a task force. We headed north.

Construction and repair crews poured in from all over the Southwestern Bell area The job was very efficiently organized. We were given the usual speech: Safety first, watch out for hot electrical wires, now get to work. My crews busied themselves with replacing broken poles and stringing new cable. We blitzed the area. Within a few days we were through and headed for home.

A storm of a greater impact hit Northwestern Bell. An Arctic winter blast roared across the flat plains of

Nebraska and Iowa. Ice and wind toppled our toll lines. When a call for aid goes out from one Bell Operating Company to another, you know the damage is extensive.

Away we went. Our first stop was West Point, Nebraska. Nights we camped in an old hotel. The rooms were large and the beds were as comfortable as most, but the hotel did have one unique feature.

The bathroom was in one corner, right out in the open. There was no partition. Heck, there wasn't even a privacy curtain to hide behind. Commode and the whole works were in plain sight.

Since I was in a private room, this was no big deal. For those who had to share a room, it must have been a bit uncomfortable. Hardly the place to take a timid bride on a honeymoon.

From Nebraska, we headed east into Iowa and soon landed at Pocahontas. This geographical chunk of America is about as flat as a dance floor. There was nary a tree to prevent our east/west toll line from catching the full impact of the storm and so it did. In our favor was the fact that said toll line did parallel the highway instead of meandering along beside a railroad. We froze our tails off, soon finished the job and headed south.

In September of 1967, we had another hurricane. It hit south of the border. Oklahoma, Texas, that is. This time it was Beulah who came a calling on Texas.

Even as we had when Carla made her destructive visit to south Texas, we left Oklahoma City in a convoy, on a route that would take us through San Antonio. I was in charge of four two-man construction crews and was reporting to my old boss from Shawnee, Mr. Lloyd Pierce, who had been my first supervisor after being promoted to management.

Our final destination was the Sun Valley of the Rio Grande. Reporters snapped a few pictures before we left the storeroom, and that was that.

After spending the day driving through numerous rain showers, we spent the night in San Antonio. Next morning it was decided that we would no longer travel in a convoy; instead, each supervisor would be responsible for his crews arrival at their destination with all possible speed. On this trip no thought was given to circling the storm or waiting it out; we headed into its fury.

The hurricane winds had diminished to a breeze, but what we encountered were torrential rains. We had scarcely begun our journey when they hit.

Behind me came eight men in four construction trucks. Also, one truck pulled a cable trailer. I had a map open on the front seat, hoping it would lead us to the town of Harlingen. And then pounding rains came, as though Mother Nature was saying, "Don't fool with me, boy!"

Visibility was limited to only a few feet as we motored cautiously down two-lane highways, moving ever deeper into Texas. We crept through towns where water was running curb deep. After several hours of this tortuous traveling, a terrible thought penetrated my thick skull: None of the towns had electricity. Soon all trucks would be low on fuel; towns without power can't pump gas. This was not a happy thought.

There was nothing to do but plod on and hope for the best. Our luck held until late afternoon. We were on the last leg of our journey when a truck ran out of gas.

We pulled winch line from another truck and hooked on. A few miles farther down the road, another sputtered and died. Now we had to pull a truck that pulled a cable trailer. Two trucks pulling and two being pulled. We were stretched as thin as possible. Eventually we limped into Harlingen on fumes and a prayer.

This country is about as flat as a calm ocean and lies only a few feet above sea level. As we pulled into town, we spotted a filling station built upon a mound that put it above the flood but not by much. There was no electricity but a tanker was parked under the canopy. We pulled in and filled all vehicles.

From here, we checked into our motel. By some miracle, it had electricity. It was too late in the day to start work.

Morning dawned clear; the water began to recede. Once more, even as it had been after Carla, we were blessed with fine weather. Nothing could go wrong now.

Or could it?

As we were becoming too smug, the Company's version of Paul Revere galloped out in his Company car and spread the warning. "The water is coming! The water is coming!" A flood was headed our way. Gather your troops and head for the motel. Pack up and get out!

It had taken that long for water deposited far upstream to reach Harlingen. By the time we pulled into the motel parking lot, flood waters were several inches deep. We were soon packed, loaded up and headed out. During this time, I had been told to take my troops several miles out of town where we would bunk in a military academy for boys.

We spent a few nights in this makeshift motel. Caterers opened up a chow line in the parking lot for both breakfast and supper. Each man was given a sack lunch. Although hardly the best of conditions, under the circumstances it was quite adequate.

It was reported that the flood waters rose to a height of some six feet in the motel we'd left. Lloyd found another that had taken on only a foot or two. When the flood waters began to recede, we moved in.

Lloyd decided he would take his supervisors to dinner in Mexico. We drove to Brownsville, parked our company cars, and rode a cab across the river, into Matamoras. He had heard of a restaurant that catered to Americans, named The Drive-In. This was a misnomer. A drive-in, it was not.

We had time to walk around and see the town. I was shocked by the poverty of the people. It must be especially difficult on the citizens of border towns to see such comparative wealth across the river. Like looking out from prison walls. Small wonder they're willing to risk so much to reach the other side. In Matamoras, most of the folks seemed to be dirt poor. A small minority were rich. There was no middle class to be seen.

The poor lived in dingy, unpainted houses that bordered the sidewalks. Dingy? Yes, but clean.

Many of these homes had windows open. People sat just inside, staring out, reminding me of prisoners or even, God forbid, animals in a zoo. Not that they were animals, but they stared at strangers in their midst with the same dull curiosity.

As we strolled along, I felt guilty for having so much while these folks had so little. They had to be content with little more than survival. Not a happy thought.

Occasionally we would pass a large home that was brightly painted and protected by a tall fence and locked gate. Here lived the rich folks who locked themselves in even as they locked others out. It was time to eat and so we turned back.

The Drive-In was a fine restaurant. Since they catered mostly to tourists, they offered a wide variety of American food, as well as Mexican. Musicians with guitars wandered about, strumming and singing.

The place had a *south of the border* atmosphere. All that dampened my spirits was the recent memory of so many destitute people. I couldn't help but wonder what they were eating. We enjoyed an excellent meal for less than it would have cost on our side of the river.

When word got out that we had been to Mexico, members of my crew asked me to take them. Before returning home, I decided to do so.

I borrowed another company car and headed south for Brownsville. We crossed the river in a bus and then, even as I had done before, we decided to look around town before eating. Some of the men wanted to spend their time and money shopping; I was content to walk the sidewalks, seeing once more the poor people and absorbing the odors of fattening foods.

Later, as we returned to the United States, we were routinely stopped by a U. S. Customs Agent who asked if we had bought anything in Mexico. Some had a carton of cigarettes, others showed him a bottle of booze. No problem with either.

One of my younger linemen, Bill, unknown to me, had bought a switchblade knife. No particular reason. I mean, he had no intention of starting a life of crime. He had simply seen the item in a store and had decided to make the purchase.

Bill didn't try to hide what he had bought. He whipped the knife out of his pocket, proud as could be, and handed it over. The agent flicked the blade open and brought it down across the edge of a counter.

He offered the handle to a startled Bill and bid us a good night. The look on Bill's face was one of pure astonishment. He learned that it is illegal to own or possess a switchblade knife within these United States.

The Sun Valley was resplendent with citrus groves and palm trees. The natives had been dealt a stunning blow by Beulah.

When water threatened our motel, we moved out. In spite of some little adversity, we were protected by Ma's umbrella. Many of these folks had lost everything. They were more cordial than I would have been under similar circumstances.

Unlike Matamoras, many of the citizens who lived in Harlingen were middle class. There were a number of Mexicans, also. No doubt but what many were illegal. One look across the river and it became obvious why these folks would go to any extreme to cross over.

Although they were dirt poor by our standards, those who had entered the United States, by whatever means, were by far better off financially than they would have been had they stayed south of the river.

As earlier stated, the land around Harlingen is very flat. There weren't enough hills and valleys to hold a lake. No water skiing, right? Wrong!

Instead of lakes, they had irrigation canals. Hardpan roads paralleled these waterways. It was a strange sight to see a car kicking up dust, a taut rope angling out to the water and held by a person on skis.

Going home the weather was good. We could buy gasoline in any town we wanted. No centipede attacked Dixie while she awaited my return.

TWELVE

INTO THE FIRE

Melvin Taylor, the second-line maintenance supervisor for the North District in Oklahoma City invited me to become one of his installation foremen. Whether or not my title had already changed as it had when Comer called, I did not know. Nor did I much care. Before he could back out, I said, "Yes, yes, yes!"

Shows how mixed up a man can get, doesn't it?

Melvin had been Comer's main man at Shawnee, that is, his wire chief. He had a passion for superb service and healthy indexes. For sure though Melvin wasn't Comer. He preferred a softer approach. More often than not, he was wearing a smile. If the smile wasn't there, beware. The reason usually became clear.

Melvin reported to Bob Baughman. Although Mr. Baughman's title was the same as Comer's, he was even more different from my former boss than was Melvin.

I reported to a satellite work center located some five miles from the district office. Tim Kubiak was firmly ensconced as cable repair foreman.

I had charge of the storeroom but Tim accepted the responsibility of keeping the work center clean. Also, he accounted for every gallon of gasoline dumped into or pumped out of the company tank.

When I arrived bright and early one Monday to take charge of installation, I found myself responsible for twelve installers and one stockman. This was a gosh awful plenty, but manageable. Had I known what the next five years held, I would have probably jumped from an overpass, taking dead aim at a spot just ahead of a Mack truck.

My crew was a good mixture of old-timers and a few youngins. Jay Worden, with whom I had worked as a lineman, was one of my older and more dependable men. Jay had an infectious laugh that would brighten my day.

The fact that we had spent many years working together as pole buddies didn't create any problems. I thought of us as friends with common goals, not any boss-subordinate relationship. Hopefully, Jay felt the same way. Jay's oldest son, Mickey, another good man, worked for me in the storeroom while attending college.

I usually had more installers working than trucks available. This created problems. Since some members of my crew worked Saturdays, usually one man was off each week day. With vacations and sick leave and all, the Big Cheeses figured we would always be able to find some means of transportation.

Even if we could find a vehicle that would run, this was hardly sufficient. These same Big Cheeses didn't seem to realize it took more than transportation if a man was to earn his pay. It required a truck stocked with material, telephones, tools, ladders, and all that stuff.

Folks with authority drove the district to see that all of our vehicles were rolling each week day. It just plumb broke old Ma's heart to find a truck sitting idle.

As supervisors, we were constantly exhorting the troops to keep their trucks clean. Store tools and telephones in the proper place. Most made an honest attempt to do so.

Then along would come some uncaring kid who, ticked off because he had no company van to call his own, invariably left the borrowed truck in a mess. Tools might come up missing, telephone inventory wouldn't balance, and there were about a dozen other reasons for not sharing. Those with trucks assigned to them felt frustrated; those who didn't have a truck to call their own felt ostracized.

We worked all residence and small business orders in five central-office exchange areas, putting phones in and taking phones out. We installed pay phones and companion booths or shelves, and even established many of the special circuits. We were responsible for seeing to it that every order was completed on its due-date, held accountable for the maintenance of our battered motor vehicles, the actions and safety of workers, employee valuations, safety meetings, quality inspections, frequent on-the-job supervision, and keeping our boss happy enough that he would direct his attention elsewhere.

Didn't take me long to wonder what I'd gotten myself into.

Ma figured that as the work load increased you should rise to the challenge. She didn't want to hire more folks or promote new supervisors or build more work centers or order more vehicles. She figured increased productivity equaled commitments met and customer satisfaction.

At least that's how Ma rationalized her actions. Or lack thereof.

We received monthly printouts showing where we stood on that all important paper totem pole. A supervisor at or near the bottom on his indexes would receive personal attention from his boss, or, if the situation was extremely serious, from those higher up the hierarchical ladder. When these folks came a calling, their typical opening statement, tongue in cheek, although a sadistic grin might cover their face, was, "I came to help." Their help was about as welcome as rain during a flood.

Our installation territory included Nichols Hills and Quail Creek, both neighborhoods housing the well-to-do along with a few families who wished they were well to do and lived well beyond their means. There was old money in Nichols Hills. Most natives were accommodating and even friendly toward us working folks.

Mr. and Mrs. Goldman were friendly folks who lived in a fine Nichols Hills house. One day, Marvin Cogburn, a veteran installer, had a job at the Goldman home. When he drove in that evening, he informed me that there was more work than what the order reflected and Mrs. Goldman asked him to return the following morning. Marvin asked that I accompany him. He didn't often ask so I adjusted my next day's schedule.

It was to be a pleasant experience. Mr. Goldman, a friendly man, was preparing to depart on a business trip. He welcomed us and then took off. We surveyed the job and decided what work needed to be done.

Mrs. Goldman was a dear lady. Much of the work was to be done in her bedroom and even more in her dressing room.

I was admiring a gold-plated shopping cart that was some smaller than those we normally push down aisles when Mrs. Goldman happened by. She stopped and asked if I knew why it was coated with gold. My first thought was to associate it with their name.

"My husband has the patent on the shopping cart," she said proudly, and danged if he didn't.

A while later she passed through again and said, "Mr. Bross, does your wife polish your shoes?"

Her question caught me off guard. I said, "No ma'am. She has four kids to worry about and plenty of her own work to do. I polish my own."

"I always polish Mr. Goldman's shoes," she said with pride.

After lunch, we were once more working in Mrs. Goldman's dressing room. She was in her bedroom next door, powdering her nose or some such thing, with the connecting door closed. Her voice came through. "Mr. Bross, a dear friend is in the hospital. Would you be so kind as to call and get the room number?"

She furnished the names of friend and hospital. I made the call. When I relayed the information through the closed door, she said, "Thank you, Mr. Bross."

A few moments passed and then her voice once more came through, saying, "Mr. Bross, I have a doctor's appointment at three. Would you please call and tell them that I will be late?"

She provided the doctor's name; I made the call and passed along Mrs. Goldman's message.

The receptionist replied, "She has never been on time before; we didn't expect her to be today."

I didn't include those words when informing Mrs. Goldman of mission accomplished. She thanked me very much. I rather enjoyed the day.

We worked in the homes of the wealthy, the very poor, and of all those in between. Although most of the men had designated routes, on any given day, as dictated by the work load, they might have to swing from one extreme to the other. Even though we didn't work attics or crawl under houses, the work can be dirty. Much of the work was outdoors, so the guys would sweat in the summer and freeze in the winter.

Having been there, I could sympathize.

Telephone inventory was an item we had to watch. We made an effort to account for every set. Even so, we had to write off many as lost. Some of these were paper errors, but it was not uncommon for our customers to take them along when they moved. Especially if they were moving out of the city or out of state. One day when our General Plant Manager, Mr. Bowen, was visiting, I asked why we didn't prosecute people for stealing our telephones. He replied that public perception was the reason. It would look as though the large corporation was going after the little guy. This made sense but it sure enough cost us a lot of phones.

He said something else that day that caused me to wonder about the man's sobriety. He predicted that in the years to come, and not that far off, we would no longer be in the telephone installation business, that we would be a wire company only. I knew that day would never arrive. Some ten years later I knew different. The man had indeed made an accurate prophecy.

Oklahoma City North District was suffering from growing pains. Those folks who were building new homes were pocketing the money. Houses had to be prewired and outdoor wires buried. Apartment houses were springing up, blooming like flowers in spring.

As our work load grew, so did the size of my crew. The day came when I found myself with eighteen installers and two stockmen. The time had come to look for that Mack truck.

I found myself working six days a week and even coming in on Sunday afternoons to pay bills and to catch up on the ever increasing load of paperwork. Even so, there was no way that I could stay on top of events that were changing so rapidly.

I wasn't alone. Ma rationalized by saying that we should work smarter. Maybe so. There is always room for improvement. Some room, but not always enough.

Ma kept her finger on my pulse. Just before I collapsed, she relented and split my crew.

Another installation crew was formed and another supervisor was brought in. We took in more territory, leaving us each with nine or ten installers. The first to come was Ollie Taylor.

Ollie had been in the district for a short time as an installation foreman, located at the District Office. We'd had little contact. I had seen him in meetings and was not impressed. I thought he was loud, belligerent, and close to being obnoxious.

Wrong again. In time we became close friends. As friends so often do, we've unfortunately drifted apart.

One morning I returned to the office and found Ollie at his desk, filling out a lengthy tool requisition, one page nearly full and ready to start another. Although it seems incredible, a first-line supervisor, or a stockman reporting to him, could order ten thousand dollars worth of equipment or material to be used for installation and never an eyebrow was raised. Other than for telephones, very little had to be accounted for.

But every blessed tool requisition, even for one single thirty cent screwdriver, required approval from a second-line supervisor. I suppose it made sense to somebody, but it is a company policy that was never explained to my satisfaction.

Anyway, as I eyed his requisition I said, "Ollie, there's no way you're going to get that past the boss."

Still writing, he said, "I'll bet breakfast that I get every item."

"You're on," said I.

We drove over to the District Office and entered through the back door. Down the hall came our boss, Charlie Johnson, a cigar-chewing man with a friendly smile. Ollie offered his list and said, "Boss, here's a few tools I need."

The previous week our Division Superintendent had made his annual visit to our work center. After his long speech, meant to inspire the troops, he wanted to inspect my tool locker. Charlie had been a witness.

Of course the Division man had to find fault. He would hunt until he did. Finally he concluded that I kept too many six-inch screwdrivers.

Charlie studied Ollie's list, chewed his cigar, and wrinkled his brow. He said, "Bross, don't you have some extra six-inch screwdrivers?"

"I sure do, boss, and Ollie can have as many as he wants."

As Charlie scratched that item, I said, "Ollie, let's go eat breakfast."

A stunned Ollie said, "Judas, I can't believe you would do that to me."

He bought breakfast. Now and then he called me Judas.

All in good fun. I think.

Sure, it was.

Eventually Ollie moved on and was replaced by a college kid who didn't like to get up early in the morning and so after a few weeks decided to take his degree to a company where the hours were better. Next came Ron Dehne and after that, Vance Duncan.

At times Ron did the unexpected. One Monday he just didn't show up for work.

There was an installer who normally loaded out when Ron was absent. I built my service orders loads and then worked on Ron's until his man arrived.

"Did Ron say anything to you about being gone today?" I asked.

"Not to me, he didn't."

When I phoned the morning report in, I asked Nancy Payne, "Did Ron say anything to you about being gone today?"

"Not to me, he didn't," said Nancy.

That afternoon I was in my office while an auditor examined my company checkbook to see that there had been no misuse of funds. The phone rang. It was Ron.

Laughing, he said, "Did you miss me?"

"Un-huh."

"I decided to take a vacation this week."

I felt the urge to kill or at least maim. "Thanks for telling me," I said, and hung up.

Ron had been raised in the far north country. Maybe that's the way they do up there, I don't know. Anyway, we remained friends. Ron moved on.

Vance Duncan came from the South District. It seems as though he'd had a disagreement with his third-line. This wasn't headline news. Most folks seemed to disagree with the man; it's just that few had the courage to speak back.

Vance had experience in telephone installation and seemed to fit into our organization quite well.

We were both reporting to Harvey Faught. As an installation foreman, I had five different supervisors. How much I was responsible for these rapid changes in personnel can only be subject to conjecture.

Before taking that second step up the ladder, old Harvey had been an installation foreman in the south district. The grapevine reported that he was a stickler for following the BSP's. The grapevine was correct.

Vance's territory was suffering from growing pains. Apartment houses were springing up in pasture lands, blooming like dandelions in the spring. They were even more troublesome.

We were under standing orders to give the customer service, no matter what. In one complex, tenants had moved in before their cable was spliced. Vance's installers laid wires across a travelled street in order to provide service.

One day, Vance, Harvey and myself were out inspecting our territory. Harvey spotted this wad of wires and came close to having a stroke.

I felt empathy for Vance as he fielded Harvey's verbal jabs. It was one of those situations where you're darned if you do and damned if you don't.

Ironically, Ollie, Ron and Vance all eventually migrated to the Company training center in Oklahoma City, teaching pole-climbing and technical courses.

In 1971, contracts between the Communication Workers of America and the Bell Companies expired. Negotiations bogged down. A nationwide strike was soon called. Supervisors were forced to man the battle stations.

We who'd had experience at climbing poles and working as installers or repairmen, did the outside work. Repair was a top priority. Only after we had this under control did we install telephones.

The Union members, of course, set up pickets outside our work centers. It was July hot. After a day or two they tired of walking in the summer sun. So, they improvised.

Near the gate they sat in lawn chairs, were shaded by an umbrella and had an ice-chest close at hand. They seemed to feel that a picket sign propped against a fence was every bit as effective as one carried to and fro.

When I would drive in, they would chide me about climbing poles in the heat and ask if I wanted a beer. I did, but I couldn't. It was all in good fun.

I think.

Before long the strike ended and all the folks returned to work. All but New York Bell, that is. More about that later. After all, we still had indexes to make.

THIRTEEN

INDEXES AND HIGH FLIERS

You know about indexes. The government is constantly tooting its own horn while flaunting indexes that prove beyond a shadow of a doubt that we're all prospering, even beyond our wildest expectations. If the country sinks, it will not be the fault of the pencil pushers in power who feed us tainted data from computer printouts. When indexing cost of living, not all items are included. Actual cost of living might go up ten percent while the indexes show it is only two.

Very deceptive!

Sure, inflation's under control. The party in the White House, whoever is in power, continues to tell us so. Why is it only those items I'm buying that shoot up in price? Due to a malfunction in the computers, no doubt. We no longer have human error. No, we most certainly do not. The blame for stupid errors falls on the poor, maligned electronic information gadgets that haven't been programmed to shift the blame back to humans or to another electronic gadget.

Now, if a certain government index rises when it should fall, or falls when it is time to rise (Yes, some indexes show improvement by rising, others by falling, thus keeping the electorate confused), then it is CYOT time. It's time to 'cover your own tail' like a yapping coyote and let others take the rap.

After all, careers in government are constantly being made and lost on the selling of how well you're doing on indexes. So too it was with Ma Bell. Ma's indexes, no doubt designed by men (and women) with good intentions have been blessed, cursed, padded, stretched, manipulated, peeled and concealed. Yes, they have.

Indexes were designed to measure the cost of labor, office space, motor vehicles, equipment, and a host of other expenditures. Supposedly they would reflect customer service and satisfaction. Sometimes they did.

Oh, the games people play.

Yes, all enterprises must have some method for measuring performances. Profits spell survival. Governments excepted, of course. They start the presses and keep printing money.

The proprietors of the "ma and pa" chicken cafe might simply study their bank statement or cash on hand. If the balance has increased, there is no need for panic. If the ink is red, sound the alarm.

"Maybe you oughta put on clean overalls every day, pa."

The reason for combining indexes and high fliers is that the two are intertwined. Back in the olden days, when I was a lineman or installer of telephones, indexes existed but were seldom mentioned to the craftsman.

We knew that if we didn't work our tails off and keep our customers happy, we would have more trouble than we wanted and could possibly be told to hit the road. I know now that Tommy Morgan must have felt the heat, but for the most part he rubbed salve on the wounds and kept on plugging. During the years, this policy changed.

In the last days and final years of Ma's glory, the responsibility for maximum performance filtered down until it reached the craftsman. We were told that any man (or woman) who failed to meet certain criterion, that is meet certain paper indexes, should be warned, then, if these paper performances didn't improve, they could and would be terminated.

The Union said, "Oh, yeah?"

Their contention was that if any dues paying member of their organization didn't perform it was because he/she hadn't received adequate training or supervision. Since the first-line supervisors didn't have a union, the buck stopped there.

As the years passed, the High Fliers, or Jets, as they were sometimes called, began to make substantial inroads within the company. My first contact with this new breed came while working as a lineman for Smiley. Jets were rare then but they did exist.

It was a warm, spring day and we were doing maintenance work on a little telephone line outside of Oklahoma City. We were on an isolated road where there was very little traffic. Come noon we ate our lunch and while Smiley sat in the truck, head bowed, enjoying a midday snooze, we decided not to disturb his dreams. Instead, we would entertain ourselves by pitching washers.

It's like pitching pennies except you throw flat washers. The winner is whoever comes nearest to a line scratched on the ground.

We had exceeded our half-hour lunch period when here came the District Construction Foreman, a cigar smoking, well liked, big Indian (some Indians did work for the Company), a man who had an ability to control his temper that was quite remarkable. A young man occupied the passenger seat.

We were seeing our first High Flier. A college grad, he was expected to absorb knowledge like a sponge absorbs water, by being exposed a few days each to various departments of the company. Given this rudimentary knowledge, once he had been adequately brainwashed and processed, he would more than likely take control of some department or other. Eventually he might move even higher.

Big Indian crawled out of his Company car and frowned at four nervous linemen. He opened the truck door and Smiley, wearing dark glasses and hat pulled down over his eyes, almost fell out. Needless to say, we terminated the washer game and within seconds were as busy as Noah when he first heard thunder.

What words were exchanged between the men I can only guess. I doubt that they discussed possible promotions or any such trivial matters.

Next day, the Jet started on the job training. He was to observe us for a couple of days and learn all he needed to know about the Construction department.

Although our initial reaction was dislike and mistrust, when he climbed into the back of the truck and made himself at home, our animosity began to dissipate. He had a way of getting along with folks, that's for sure.

On the way to the job, he said, "When we drove up on you yesterday, I thought big Indian was going to have a heart attack. After nearly biting his cigar in two, he said, 'Ho! Ho! Ho! Foreman's asleep in the cab and the boys are pitching washers. Ho! Ho!' I doubt though that he was all that amused."

I doubt it, too.

Yeah, the Jet won us over. He didn't stand back and observe from afar but seemed anxious to pitch in. He knew the way to learn was to dip your hands into the muck. Before that first day ended, he was even an elementary pole climber. What happened to the man after he left us, I have no idea. He could have quit the Company and gone elsewhere or he could be the head over one of the so-called Baby Bell Companies for all I know.

Once the Jets had a toehold they began to wrest control from the old hands who had fought the wars. These were college grads, individuals at or near the top of their class, white male (No women were considered for outside jobs at that time), fast climbing folks whose goal was to reach the top of the heap even though the higher they climbed the fewer the opportunities.

As the years passed, the Jets became as thick as buffalo chips on the prairies in days of old. Some folks thought they were far less useful.

Not all of these wonder kids survived. They were hired into first-line management, promoted promptly to second-line, and then, if they were worthy, moved to that coveted third-level. The Company's policy was if a Jet wasn't capable of reaching third-line in five years, out they went. Many were culled. If their indexes were all headed in the right direction, doors were opened. If not, there was no job security.

During the time I was an installation foreman, there was an opening for a supervisor on the other crew at my work center. One day in early summer here came my boss with a youth who had recently graduated from one of our State Universities. On this, his first day on the Company payroll, he and I carried the same title.

It was mid afternoon. The boss man asked me to teach this neophyte all he needed to know and then old boss hurried away. Along about four, the kid decided he needed to leave for the day. In the tone of voice a son might use towards his father, he asked, "What time do I need to be here tomorrow morning?"

He wouldn't last. For sure he was smart, well educated, and a whiz at the books. However, for a Jet to have success within the Bell System, other qualities were needed. It helped to be aggressive, overbearing, haughty, contemptuous of subordinates, and, of course, arrogant.

Jets had to fight for that next rung on the ladder. If it meant knocking others off, so be it. The meek might inherit the earth, but they would never run old Ma Bell.

As he awaited my answer, I felt sorry for this chubby young man who was already fighting (and losing) the battle of the bulge. "That's up to you," I said. "I'm usually here between six and six-thirty."

Eyes opened wide and mouth flew open. I would not have been surprised had he dropped to the floor in a frothy fit.

"I thought this was an eight to five job," he said.

It seems as though certain information had been omitted from this recruit's indoctrination. "If you can manage your job from eight to five, I want to take lessons," I said. "However, if you want your crew to roll out at eight, you'll need to be here in time to build each man a load of orders and then get the equipment sheets to the storeroom. Also, you need to phone your morning report to the District Office before that time.

A very sad and disillusioned college grad left the work center that afternoon. He was so down that I even wondered if he would show the next day.

He did. Barely. The next morning he arrived about seven, still yawning. I guided him through the routine. He only lasted a few weeks and then offered his degree to a company whose employees kept decent hours. Had he held on for several months and developed a mental toughness, he would have eventually been promoted to a job where the hours were better.

Mr. Baughman was not a Jet. He had worked his way up through the Company and his District usually enjoyed performances that were right up there near the top. Consequently, he was promoted.

A succession of High Fliers followed and our performances began to slip. Perhaps this was only a coincidence, but during the next few years we went from being first place to being near the bottom.

Having very little if any personal experience on which to draw, the Jets had to rely upon college and Company training. In both, there is a certain amount of role playing designed to prepare the trainee for any conceivable situation they might encounter. Trouble is, there are too many inconceivable situations that pop up.

Historically, the first working day of a new year is one of our heavier days in installation. End of the school semester and a new calendar hanging on the wall causes numerous folks to move from town to town or to another neighborhood. We have taken time out for two holidays and so our work load has accumulated.

Adding to our misery, Mother Nature, on one such day, became angry for reasons known only to her and delivered an enormous ice storm. I was in my office before six that morning, built my installation loads and called a service station to send men to chain the truck tires. They had only one man to send.

As soon as possible, I helped. By eight, most of the trucks were ready to roll.

Even with chains on, the men had to drive with extreme caution. Walking was even more hazardous. At day's end, every man returned with nary an accident to report. This in itself was a minor miracle.

We missed one due date service order, one that called for installing a line for a small business. Even then the installer arrived on premise at six that afternoon and found the place closed. It's doubtful that anyone had even come in that day to open the business.

All in all, I felt good about our performance under such trying conditions.

The next day, when phoning in the report, I had to list a Company missed order since it was a 'no access' after five P.M. Later that morning, the District man, a young, rising Jet, called and said, "Why did you miss that order yesterday?"

I thought he was joking. He wasn't. When I tried to explain why, he said, "I don't want excuses; I want results."

This saying, trite though it may be, was a favorite of many Jets. These words were also parroted by many of the up through the ranks, second-line supervisors, or those who hoped to be.

"We did have an ice storm," I reminded him.

"That's not even a good excuse." Voice rising, he plowed ahead. "A man can work as many orders in bad weather as he can in good. If you want to work for me, don't miss any more orders!"

What do you say to a man who never in his life has driven a company truck on ice, climbed a slick pole, skated along on glass sidewalks, or, for that matter, ever installed a phone in good weather? He had no doubt spend the previous day at his desk, studying indexes and reports.

Some of these guys are book smart, but come up a little short in the common sense category. What griped me was he slammed down his telephone handset a split second before the idea struck me.

Jets can be sneaky fast.

Another day I received a telephone call from a Business Office Representative saying that one of my installers had removed telephones that were no longer in use from a vacant house. There was nothing wrong in doing so, other than the owner wanted to know how the man had gained access.

I asked the installer. He informed me that he had gone by at noon, the painters had been working inside, and they had left the house unlocked while at lunch. My man had merely walked in, disconnected the phones and left.

Sounded logical to me. I phoned the customer. After I had repeated the installer's story, he said, "Mr. Bross, it didn't happen that way. The painters assured me that they locked the house before going to lunch."

Well now, this was a new wrinkle. It made me wonder how my man did get in. "Did you find any sign of forced entry?" I asked.

"No, that's not the issue. I know that the former tenant had keys made and I suspect that he gave one to a neighbor. It is important that I know who let your man in. Otherwise, I will have to change the locks."

"Had it happened that way, I see no reason why my installer would have hesitated to tell me. He said the door was unlocked."

Out of the blue, the man shouted, "Mr. Bross, your installer is a damned liar!"

Don't lose your tempter. Count to ten. We know the customer is always right. I lost my temper.

"I don't have to listen to this crap," I said, and hung up the phone.

Some half hour later, after having received a personal invitation to come for a visit, I was in the District Office, seated across from the current Jet, a pipe smoking, foot shuffler, who usually answered his subordinates with a grunt. Our customer, of course, had called the Business Office and lodged a complaint against yours truly. Said complaint had been promptly forwarded.

The Jet wanted to hear my version. After I told him, he gave me his sternest frown, the one he always thought intimidated folks, and said, "We don't hang up on customers."

"We do when a customer calls my installer a damned liar," I said.

"Not even then," said he.

Well, I'm sure that was a black mark on my record. Seems as though I did have to listen to that crap after all. Although I never admitted to the Jet that he was right, he was. We shouldn't ever hang up on a customer. Still, we're only human.

During this same Jet's tenure in our District, my title was changed to PBX installation foreman. For a time my office, along with that of Frank Appl's, the other PBX supervisor, was in the same building as the district office.

One day Paul Taylor, our second-line, said to Frank and me, "The boss has been complaining about you two."

"What's his problem?" I asked.

"He walked by on his way home yesterday, about five-thirty or so, and you were both gone. He doesn't think it's right that you should leave before he does."

"Most days I'm here later than that," I said.

"One or both of you should be here every day until after the boss leaves."

By now I was becoming a little irritated with this line of thinking "What time does he come to work in the morning?" asked motor-mouth me.

"That has nothing to do with it."

"That has everything to do with it. Frank is here by six; I come in before six thirty. Your boss usually arrives about eight or maybe a little before. If he'll join us in the early morning, I'll make it a point to stay until after he leaves in the evening."

Paul turned away, no doubt wondering if I had been insubordinate.

Whenever a Jet came to our District as a first-line supervisor, we who had worked our way up one pitiful step to that level, knew that if he held on he would be given preference in any promotion. Still, you grit your teeth and help all you can. Chances are that some day he will be your boss.

I'm sure it sounds as though I believe the High Fliers were the ruination of Ma Bell. No, I am willing to look at the positive side.

And there is a positive side.

For the most part, old timers resisted change of any sort. With some few exceptions (and there were some), they were satisfied with maintaining the status quo. They had a Henry Ford complex; a black telephone was good enough for anyone.

New blood brought changes, many of them for the good. The Jets were willing to experiment. If their current plan failed, they would back up and plow in from a different direction.

New equipment and services were offered. Our gray trucks gave way to bright colors. These men (and women) saw the need for dialogue between the various departments and made it happen. Increased profits became a primary goal. Being realistic, we know that a Company without adequate profits will falter and die.

When the U. S. Government said we had to have a quota of blacks and women in the outside departments, the Jets worked towards this objective. Many of the old timers were appalled. The new people made sure that it worked.

Their time had come. Although I got crossways with some of these boy wonders on several occasions, I know now that the Company could never have survived without change.

There: It was hard, but I said it!

As for promotions into second-line, these were not all handed to the Jets. Some who had started as a craftsman, even as I had, reached this level. Why didn't I?

Give me enough time and I'll come up with an excuse or two. But for now it's time to return to the strike of 1971.

FOURTEEN

A BITE OF THE APPLE

Although a majority of the Union members voted to ratify the new contract negotiated in 1971, Ma's New York children said, "No way!" While folks in other Bell Companies returned to work, these stubborn Yankees chose to continue the strike, to stand alone, if necessary, in order to gain further concessions.

Mother Bell responded quickly, as though a storm had struck. She brought folks in from across the U. S. of America to pitch in and keep the system humming. Since craftsmen everywhere were Union brothers and sisters to those striking, she couldn't ask for or expect their help. So, she turned to her next line of defense.

Ma asked for volunteers from her ranks of first-line supervisors, specifically those with some experience in climbing poles and repairing telephone instruments. Recruits must agree to stay in New York a minimum of three months, should the strike last that long.

I was volunteered.

Charlie Johnson, my boss at the time, called me into his office. "Do you want to go to New York?" was the question he asked.

"I don't know. Let me think it over."

"You've thought long enough. You're going."

Heck, I was ready for a change anyway.

For a country boy who had never been nearer to the Mississippi River than Jonesboro, Arkansas, it was a memorable experience. I was soon aboard one of American Airline's 727's, heading for faraway places.

Since we neither crashed nor detoured to Cuba, the plane touched down safely at Kennedy Airport. Was I apprehensive? You bet. When I entered Kennedy, with all its international terminals, I might as well have been lost in Bangladesh. And where were those folks from New York Bell who were supposed to meet and greet me? Try as I might, I couldn't see anyone who might be a child of Ma Bell's.

They found me at the luggage claim, waiting on my suitcase to appear. How they picked me out of the multitude, I'll never know. Perhaps they had been told to keep an eye out for a bewildered, disoriented, country boy who looked like a lost Okie. Whatever, I was soon loaded into a van, along with several other confused souls, and driven to Westchester County.

Darkness descended before we reached the motel, a nice enough two-story habitat located near the Interstate Highway between Peekskill and Yorktown Heights. The next morning I found that I was only one of many who awaited the unknown.

The New York folks treated a multitude of us foreigners to breakfast and then escorted us to the work center near Peekskill. That place was literally swarming with Yankees. Like Peekskill itself, the work center looked as though it had been there since days of old. Still, it was well arranged. Company vans were parked inside and the storeroom was fully stocked.

We wedged into a packed meeting room and were greeted by the District Superintendent, who surely must have felt as though he had been invaded. It is doubtful that he had ever seen so many rednecks and must have been somewhat apprehensive about loosing us upon his cherished customers. Could be that we would set public relations back a decade or two.

He thought it necessary to drum instructions into our skulls before exposing us to the natives. This was his right. He ran the show. Here was the Comer Williams of Westchester County. His approach, although firm, was milder than Comer's. He fixed us with a stern eye and said, "Even though you will be doing craft work, we're not about to treat you like craftsmen. We'll not forget that you are supervisors."

A man from Port Arthur, Texas, whom I had already met, was Jack Honey (pronounced Ho-ne). Jack spoke up. "Shoot, I was hoping to be treated as good as a craftsman for a change," he said, with a dry wit and a straight face.

A slight hesitation, an intimidating glare from the man up front that failed to intimidate, and then the poor soul continued. "After hours it is permissible to drive a company car to a restaurant. But, if you get drunk and wreck the car, you'll be put on an airplane and sent *all* the way home."

Jack responded by saying, "I'd sure hate to be sent just half way."

We all cracked up. All but the District man, that is.

After informing us that we would work six days out of seven and would be paid for all overtime hours (Glory be!), he seemed happy to conclude his portion of the meeting and turn us over to our second-line supervisor for further instructions.

A ball of fire named Pete was our second-line. He expressed himself clearly and forcefully, but it was soon obvious that Pete was okay. He was more than okay; he was a heck of a nice guy, even though he was a Yankee.

We were given street maps of the area, told that we would be working repair, assigned vehicles, and handed some trouble tickets. Then, we jumped into vans, fired the engines and roared out of there, as though we knew what we were doing. As Pete stood alone on the loading dock, waving a clenched fist and shouting, "Bonzai! Bonzai! Bonzai!" we invaded the north.

While the morning was still young, Westchester County in New York State was filled with a number of lost Southerners. That first day or two, we spent considerable time lost in some culdesac or parked alongside the road, studying maps, wondering where we were and how to get to where we hoped to go.

Two of my coworkers, with whom I soon became friends, hailed from Texas. Harry Hogue, a big man from Dallas, stood as tall as Marshall Dillon. Dress him in boots and a ten gallon hat, pin a star to his shirt, mount him on a horse, and the outlaws would run for cover.

The other man, from Tyler, was Jim "Spider" Webb. A lithe, wiry man, he had a quick wit and seemed at home anywhere. Spider had a special knack that eludes so many of us. Once he had ridden a route, he could drive it without any problem. Even if he hadn't ridden it, he could look at a map and take you straight to your destination. Or he could follow verbal directions.

For those of us who had trouble finding our way to the motel, it was disgusting to find one so talented.

The three of us were neighbors. By the luck of the draw I had been given a private room, with kitchenette. Jim and Harry shared the next room. Since there was a connecting door between my room and theirs, the three of us passed back and forth at will. And of course we kept various liquid refreshments in my refrigerator as well as sandwich stuff and all that.

We would often go out for supper (That's dinner to you Yankees) together. Neither me nor my Texas buddies had ever seen New York City, other than for landing there on an airplane and then we had been hurried away.

We talked Pete into letting us off work the same Wednesday and headed for the land of skyscrapers. Spider drove the company car to the train station at Croton on the Hudson (What a name), stashed it in a lot large enough to park an O. U. versus Texas football crowd, and we bought our tickets. Soon we were on our way, as excited as schoolboys on a field trip.

Around both Peekskill and Yorktown Heights, the topography is hilly, even mountainous, resplendent with tall trees. However, as we approached New York City, the land began to flatten. We travelled through towns, stopping at some and rolling through others. And then we entered Wonderland.

From the train we saw Yankee Stadium. In this arena were the footprints of The Babe, DiMaggio, Yogi, and Casey. Even more, from my viewpoint, it was the setting that had recently highlighted the multi talents of Mickey Mantle, Oklahoma's own. I regret that during my stay in Westchester County I was unable to visit the Stadium and see the team play.

The train rolled. We rumbled past weathered tenements, where so many lived with so little hope. From high above, black kids stared down from tiny balconies. To one who had been raised in the wide open spaces, the scene was incomprehensible.

For the first time ever, I began to feel empathy for these street kids and for their parents. What chance can a kid, either black or white, have in this country if his world is asphalt and concrete and crumbling tenement buildings? It's amazing that as many as do make something of themselves.

As we rushed toward the Grand Central Station I made an effort to push these views of abject poverty from my mind. They would never completely go away. First things first, though. I was in The Big Apple!

The train stopped and we stepped off. We flowed with the natives until we reached the street. Outside, we were in a new universe. Strangers in a strange land. We moved towards the heart of the city.

Not having any desire to be thrown into the pokey for a misdemeanor, we were extra cautious. When we reached an intersection and saw the flashing light that read, "Don't walk," we stopped.

It didn't work. The flow of foot traffic pushed us into the street. Horns honked. Automobiles and a host of pedestrians fought for the same inch of territory.

It seemed as though about half of the vehicles were taxicabs. Judging from the sounds we heard, most of the cabbies drove with one hand on the wheel and the other on the horn. Part of their training, I suppose.

We moved ever deeper into the heart of the city. Then, without warning, a splendid sight appeared suddenly, rising majestically towards the sky.

There was King Kong's ladder, the Empire State Building. Even the flow of pressing natives couldn't push us further. We froze in place. Harry's head lifted slowly, as his eyes moved up until he viewed the top. Then, in his slow, Texas drawl, he said, "My God, think of all the hay that place would hold!"

I doubt that any of you New Yorkers had ever thought of using the building for that purpose.

We wandered aimlessly about, absorbing the city's sounds, its odors, and its feel. We stopped for lunch at Jack Dempsey's Restaurant. Although we didn't see the great American, who at that time was still alive and was said to drop in occasionally, we did feel his presence.

At the end of the day, we hadn't scraped the tip of the iceberg. It was time to catch the train.

Although I didn't have the opportunity to become acquainted with many citizens of New York City, I was surprised to discover that natives of Westchester County differed little from most of us Okies or Texicans. The majority of those I met seemed to be decent, hard working folks who were concerned with buying bread, raising their families, digging up enough money to pay taxes, and educating their children. Their wants and needs were little different from ours. As for friendly, I encountered fewer snobs while working in Westchester County than I had while installing telephones in Oklahoma City.

One day I entered the home of an elderly woman, an Irish widow. As I worked on her telephone, she said, "Young man, would you like a cup of coffee?"

"Yes ma'am, I sure would."

She plucked two cups from her cabinet and filled each about half way with perked coffee. Then out of her cabinet, she brought forth a bottle of whiskey, poured a generous amount into her cup and, holding the mouth of the bottle over mine, said, "Tell me when to stop."

"No, thank you. I can't drink on the job."

She stared at me in disbelief, shuddered, and made a face. "I don't know how in the world you can stand to drink it straight," she said.

She had me there. I didn't know what to say.

For the most part, even folks on strike tolerated this invasion of foreigners into their home community. One day I was parked along the street, back van door open, searching for material to use on a repair. Behind me, a pickup pulled to the curb. Two men got out. I had the feeling that they were striking telephone workers and that they spelled trouble with a capital T.

I was only half right. They were on strike. It soon became clear that they didn't want trouble.

They were interested in where I was from, what I'd heard about the negotiations between Company and Union, whether or not I thought a contract would soon be reached, and that sort of thing. In order to survive, they were working at odd jobs. Had the tables been turned, I'm not sure I could have been so gracious.

During my time in Westchester County, there were only a couple of incidents that made me uneasy. The first was when I drove out into the country to work a trouble report. The customer was a small business, a service station and grocery store combined.

As I drew near, the cause of their trouble was obvious. Their service wire, from pole to building, spanned over the drive. That is, it was supposed to span, but it had been pulled in two. It is probable that some truck had hooked the wire and done the job.

I pulled the drop wire reel off the truck and soon started to work. Out of the station roared a hot headed young man, on strike against the Bell System. His view of me was that I was a strike breaker. He lashed into me verbally, calling me many uncomplimentary names.

When he paused for breath, I said, "If you want your phone fixed, get the heck off my back."

His grandmother, the proprietor, heard enough of the exchange to form an opinion. She sent the young man packing and I replaced the wire.

There was another time when I was dispatched on what was thought to be a cut cable. My mission was to find the cut, then phone in the location so that a cable repairman could be dispatched.

I travelled a lonely, desolate and narrow country road. Upon reaching the point where the cable came out of the ground and went up a pole, I climbed up to the cross connect box and opened the lid. Oh, boy!

Someone had climbed the pole ahead of me and ripped out all the cross connecting wires. In my mind there was no doubt but what it had been done by one or more militant strikers.

I phoned the test desk and reported my find. The deskman procured cable records from assignment and one by one told me what cross connect to make. Since there were some eighty or so, this was time consuming.

All the while I felt as though I was being observed from somewhere up on the mountain. Not only watched, but viewed through a scope. Mounted on a rifle.

Paranoid? Perhaps. Anyway, I finished without being fired upon and made it back alive.

Spider told of an incident that happened to him one day. He was working on a pole, near the curb. Out of the house came a young boy, four or five years of age. He stood on the sidewalk, beneath Spider (A dangerous place to be), and constantly repeated the word, "Scab."

Spider said, "Son, who told you to come out and call me that name?"

"My daddy did," replied the boy.

"I want you to go into the house and tell your daddy to come out here."

The boy hurried in and in all probability delivered Spider's message. Papa didn't show. It took a brave man to send his son on such a mission. Ah well, thankfully he wasn't typical.

There was, of course, a language barrier between us and the natives. Anytime they couldn't understand what we had said, they would invariably say, "Pardon me?"

Jack Honey's standard reply was, "Why? What did you do?"

It was Company policy that at the end of each month they would either fly us home for the weekend or, if we chose, fly our wife to us. I flew home once; Dixie flew to New York the other. She was to land at La Guardia Airport on a Friday night and my dilemma was how to transport her from there to the motel. I asked Spider to be our chauffeur.

"I'm going home for the weekend," he said. "I'll drive us to the airport and you can drive Dixie back."

"I'd probably get lost."

"Hey, man, that's not possible."

Of course Jim was able to find the airport without any trouble. Around nine that night Dixie's plane landed. She and I decided to stay until Jim left at ten. After seeing him off, we found the car. Even before leaving the parking lot, I got lost.

Although I eventually found the access road to the highway, I missed the on ramp and didn't know how to find it again. As we wandered around New York City, I had no earthly idea of how to find Westchester County. The logical solution would have been to stop and ask directions. There was only one problem: by then we were in Harlem.

Now I don't wish to disparage the good citizens of Harlem, but we were apprehensive and frightened. This was even worse than had been the Golden Pheasant those several years ago. After midnight I was alone with my wife in a New York Bell car. In an unknown section of the City whose reputation I knew only by rumor.

I believe I was more worried about damage being done to the Company vehicle that was supposed to be back in Westchester County than I was for our personal safety.

To their credit and our good fortune, no mob stormed the car. As we cruised the streets with locked doors, we did receive some menacing stares.

The Lord sometimes has mercy on fools. We did eventually find the highway. It was four o'clock in the morning before we were safely ensconced in the motel.

I took a day off work in the middle of the week and Dixie and I caught a train into New York City. With help from the Grayline bus tours, we spent the day sight-seeing. This was a good investment.

We visited St. John's Cathedral, rode down Wall Street, and stood on the observation deck of The Empire State Building. Then we rode the Staten Island Ferry out to the Statue of Liberty. This was of special significance to me since my maternal grandparents had arrived by boat into this same harbor.

We ached to see a Broadway Play. Once again, Spider was our salvation. One night he drove us into the

City. While he and Harry went to R.C.A Music Hall, Dixie and I saw the play, "Promises, Promises." We were not only fascinated with the superb acting performances, but awestruck with the rapid changes from one scene to another.

Although the strike was far from over, once my three month hitch was nearing an end, I decided to go home. Dixie returned to New York and when I wasn't at work, we spent what we could of the last week together.

It was a beautiful autumn. The trees were quite splendid in their multi colors of orange, green, red, and brown. We fell in love with Bear Mountain and the sail-boats on the Hudson. After work and on my day off, we wandered through some old cemeteries, marveling at the headstones of Revolutionary War soldiers.

Pete offered to try and get me a supervisor's job in New York if we would agree to move. Had it not been for so many family ties in Oklahoma, we could have both been content with living in this newfound paradise.

On Thanksgiving Eve, we said our good-byes, were driven to La Guardia, and boarded a plane. Our feelings were mixed. Sure, it would be good to get home again even if I was returning to my old job, one that I was quite tired of by now.

Unknown to me, a new job was not far off.

FIFTEEN

PRIMA DONNA'S

During my years as a lowly craftsman, we thought of PBX installers and repairmen as being somewhat aloof, true prima donna's. Here were men who strutted to work wearing dress slacks, short sleeved shirts, and low-cut shoes. Their attitude was smug, even perhaps somewhat haughty.

They spent time, or so we thought, primarily in offices of larger business customers, taking up residence in coffee bars, and flirting with pretty little secretaries. In the twinkling of an eye, I found myself with the title of a PBX installation foreman. With the title comes knowledge. Right? Hmmm.

Harvey Faught was my immediate supervisor; the Jet who was known for his foot shuffling was our district man. I was called into the Jet's office and informed of my promotion. Yes, I would remain a first-line supervisor, but being boss over such an elite group was considered a step up in the pecking order.

They spent a good amount of time explaining the magnitude of my new job and how great performances (high indexes) would be expected from me and my crew. Out of the kettle and into the pot I went.

Frank Appl's PBX installation crew had grown until it was no longer manageable. It was comparable to what mine had been before they'd divided the workmen and brought in Ollie. Before old Frank lost his sanity or experienced a nervous breakdown, Ma decided to give the poor man a break.

Oklahoma City, as well as the entire State, was enjoying a period of prosperity that would continue until Penn Square Bank came tumbling down, and soon after that the oil patch boom would grind to a halt. These calamities were far in the future. Even as the residential section was growing so fast that we could hardly keep up with the demand for service, so was business. Banks and oil companies were sprouting like mushrooms in a dark cellar. Folks were reaching for the big bucks.

Mercy Hospital decided to build new, out in the far suburbs. Satisfying their telephone needs kept two of my men tied up for months. We began adding additional installers to the crew.

My first morning on this new job, I felt nearly as frustrated as I had that first day in Shawnee. However, I did enjoy an advantage or two I'd not had before. By now I had considerable experience supervising and was acquainted with most of the members of my new crew. As for the work they did, I had little idea.

Frank was my salvation. He knew Ma Bell's PBX installation practices better than most anyone I'd ever met. Heck, he'd worked in Comer's District so long that he had to.

Frank taught me the best method to presurvey a job as well as how to order the necessary telephones and associated equipment. Once you know these functions, the battle is half won.

Frank and I shared a Company car. Being the junior man, I was elected to drive. One day when we were heading out to check a job and I had hit about six red lights in a row. Frank, a soft-spoken man, turned to me and said, "I've been riding with you for nearly a year now and you haven't hit a green light yet."

Sometimes it seemed that way.

Frank had told me that he was buying a new car for Helen, his wife. One morning I arrived about six-thirty and there sat a new Pontiac, the only car in the lot. I went inside and said, "Frank, how do you like Helen's new car?"

"Okay," he said, "except the darn thing don't have a dimmer switch. I drove to work on brights and must have made about a hundred people mad. I've never heard so many horns honkin'. It's a wonder I didn't kick a hole through the floorboard."

This for Frank was quite a long speech. No doubt but what he was troubled. It did seem as though those folks at General Motors, those engineers responsible for such, would have provided some method or other to dim the lights.

After our crews hit the road, we went forth to find the elusive dimmer switch. After a thorough search of the floor and finding nothing on which to step, we turned to the owner's manual.

Sure enough, Frank had bought one of the first cars with the dimmer switch on the directional light handle. I'm sure the folks at G. M. were having their little laugh as buyers all over the country experienced similar dilemmas.

A large, metal building sat at the back of the lot at our work center. This was divided into rooms, "ready rooms", as they were called, since this was where the craftsmen gathered to make ready for a day's work. Other space was used for our office.

Frank and I shared an office. The day came when the Company needed our space for an additional crew. Some of the folks upstairs decided that the best policy was to segregate supervisors from craftsmen. Our offices were moved into the main building. I had no problem with this arrangement.

Our desks were back to back. The office floor was covered with tiles, one foot square. One day I noticed that Frank was turned sideways in his chair and staring hypnotically at the floor. "What you looking at, Frank?" I asked.

"Come see," was his brief reply.

What I saw was a little black ant who had invaded our territory and now seemed hesitant to cross the crack between the floor tiles. He would run up to one edge, back off, and then run up to another.

Amused, we continued to watch his antics for several moments and then Frank stepped on the poor little fellow.

"Why did you do that, Frank?"

"I just sent him to piss-ant heaven," he said.

A supervisor for an additional installation crew needed office space. She moved in with us. Yes, she!

By now, thanks to Uncle Sam's persistence and insistence, the company not only had female installers, but female outdoor supervisors as well. Ann's desk was placed behind mine.

I'll tell you true: Frank and I had both graduated from the old school and weren't quite ready for such a radical change. Fact is, though, Ann fit right in and soon the idea of a female installation supervisor didn't cause either of us to go into convulsions. Her presence seemed natural. Tradition had been shattered, but what the heck?

One day we were all three seated at our desks and Frank and Ann were trying to carry on a conversation. Ann would mumble something and then Frank would respond with his own version of the mumble.

Ann mumbled, "Frank, way you mumble, I can't understand a word you're saying."

Frank mumbled, "The way you mumble, I can't understand a word you're saying."

As I was leaving the room I said, "You think you got problems, I can't understand either of you."

Later, when Frank and I were alone, he asked, "Do I mumble?"

"You mumble, Frank."

Evidently no one had ever told him.

The uppity-ups in the Company, in their infinite wisdom, made an about face and decided that a supervisor's office should not only be in the same building with his subordinates, but in the same room as well. Although Frank and I protested, we were forced to move our offices into the ready rooms.

By protesting, we were accused of being prima donnas. Every man to his own opinion, I suppose. More than likely, it depends on where you're sitting.

One of the installers I hardly knew upon taking charge of the crew was Bill Fielder. We established a steady rapport that led to trust and friendship. He was a crackerjack technician, a man who would work his tail off and put in overtime if necessary. He would unless it was the weekend and he had a lake outing planned. He and his wife own a waterfront lot at Lake Eufaula where they keep a mobile home.

Come a weekend or vacation time, you could bet your next to last dollar that he and Mamma Kay (Bill's wife, so tagged because she not only treated their kids to the pleasures of laking, boating, and water-skiing, but friends of their children as well), would pack up and head for this retreat. Here they were about as isolated as they could possibly be without strapping on backpacks and wandering off into the wilderness. They had no telephone in their mobile home, nor did they want one.

The Company'd had their eye on Bill for some time, realizing that he had the potential to do more than install telephones. Frank and I had both recommended him for a management job, a recommendation that has been compared to the kiss of death.

Bill couldn't care less about having a management title. He was content in all ways. Not only did he enjoy being a technician, but he had a close knit family and the perfect hideaway.

It was midweek and Bill was on vacation. There was no doubt about where he had gone. My boss called and said the company wanted to promote Bill to PBX Installation Foreman at Enid, Oklahoma. They needed an immediate answer.

My mission: Find Bill and ask him to call the District Office during the early morning of the following day. Don't tell him why he is to call, only that he is. Yeah, you bet.

How to find Bill? I put out feelers to members of my crew and learned that the Fielder's son, Gregg, had stayed home since he was working or going to school or some such thing. All I had to do was locate Gregg.

One tends to forget that young men often find distractions that prevent their heading home early in the evening. I must have worn the bells out on Bill's telephone as I constantly dialed his home number throughout the afternoon and into the evening.

Finally, about nine or so, Gregg must have come wandering in and heard the telephone. I explained that I needed to contact his pappy in the worse way.

"Why?" asked Gregg.

A natural enough question, but this was a hush-hush, top secret operation. So my answer was vague and noncommittal and so forth. In spite of the intrigue, Gregg came up with the telephone number of friends who lived only a block of so from Fielder's retreat. If only they were home....

They were home. I asked the man if he would be so kind as to have Bill phone me, collect. He would and Bill did.

Now old Bill was friendly on the phone (Old! We're roughly the same age), but he'd been boating and sipping suds and was in no mood for intrigue. It took him about fifteen seconds to sort through the garbage I was feeding him and his immediate answer was similar to, "Hell, no! I won't go."

I would probably have done Bill and Kay a favor by apologizing for having bothered them and then wishing them a happy vacation. Perhaps I wanted Bill to share my misery; I don't know. Oh, it wasn't that bad.

Anyway, I didn't hang up, but kept chipping away at Bill's defenses, exhorting him to accept the challenge, explaining that the move to Enid would be temporary and that he would soon be transferred back to the city. Bill wasn't buying it.

Still, I had the feeling that he was beginning to weaken. After more than an hour on the phone, I said, "Look, Bill, don't give me an answer tonight."

"I've already given you my answer."

"Talk it over with Mamma Kay and call me in the morning."

"There's nothing to talk over."

"Come on, Bill--sleep on it."

Reluctantly, he agreed to do so.

He didn't sleep on it or anything else.

Early the next morning my telephone rang. "You SOB," he said.

"You're going to take it, aren't you?"

"Don, we haven't even been to bed. Me and Kay have been thrashing this out all night."

"You're going to take it, aren't you?"

"Yeah, I'm gonna take it."

Bill accepted the job and they moved to Enid. The company did transfer him back to Oklahoma City, but perhaps it did take a little bit longer than I had hinted at during our marathon visit on the phone.

Dixie and I visited the Fielders and became good friends. Bill, in a joking manner (I think he was joking), would always say, "You SOB! Look what you've done."

An old-timer on my crew was Paul Williams. He was as slow as cold molasses, but Paul could install any piece of PBX equipment known to man.

One day Paul said to me, "You can just call me P.O.T.B.S." When I asked what that stood for, he said, "Pride of the Bell System." I told him that I thought the B.S. might stand for something else.

Even though we were friends, I would sometimes want to wring Paul's neck. He was stubborn and very much opinionated. Not necessarily undesirable traits unless they're carried to extremes.

Paul couldn't accept the rapid changes he felt were wrecking the company. We made one change in particular that caused him great discomfort.

Since Oklahoma City North had no female PBX installer, it was decided that one was needed. It was further decided that said female would become a part of my crew. Paul could not accept this.

Her name was Janet, and she reported bright and early one sunny morning. She was lean with a quick wit and had been a basic telephone installer in the South District. I was alone when she walked into my office that first morning and introduced herself.

"How are you?" I asked.

"Scared to death," was her brief reply. She looked scared as she settled into a chair.

Sure. Why wouldn't she be?

I wasn't pleased with breaking tradition by having a woman on my crew, but in fairness to Janet she did an admirable job in less than desirable social conditions. Other installers drifted in; most of them gave her a cold shoulder.

When Paul strolled in and saw her sitting at the ready room table, he made an abrupt about face and visited another room. At eight o'clock I had to take him his orders. He accepted them without a word and walked out of the building, a very unhappy man.

Surely within a day or two Paul's resistance would weaken. It didn't. Paul steadfastly refused to be in the same room with Janet.

If I was holding a safety meeting or some such thing, Paul would stand in the doorway and look in.

Come evening, if he happened to be sitting at the table preparing his daily time report and she walked in, he would gather up his papers and move next door. To my knowledge, until the day he died, Paul never spoke one word to Janet.

Of course I had to counsel with Paul and make an attempt to have him come around. He wouldn't, but he did tell me why he felt like he did. It came back to that old bugaboo, "Affirmative Action".

Paul's anger was directed towards Government policy. Regardless of lack of seniority or qualifications, a certain number of our positions had to be filled with various minorities. I tried to explain that like it or not, there was nothing we could do. Paul didn't buy this. He protested in the only way he could.

After several years as a PBX Installation Foreman, I become Installation Control Foreman for the Oklahoma City North District. This position was created for three reasons.

First, control center clerks worked a late shift, thus enabling them to load out all basic installation orders due the following day. Since many of these orders didn't come off of the teletype until late in the afternoon, this shift was essential.

The clerk would also allot a specific time for each work operation. If she had more work orders than she did installers, it was the responsibility of the installation foreman to find a way to get the danged things worked.

A second function for control was to monitor the work loads throughout the day and be certain that all orders were worked by their due date. A control foreman could shift any installer to meet customer commitments.

Since his title was still a first-line management entry, installation foremen resented him meddling with their troops. Above all, diplomacy was required.

Third, the control foreman was to act as liaison between installation and other Company departments. If an installer was having trouble with a person in another organization, he/she was to call the control foreman who was then responsible to clear the roadblock.

My entire crew consisted of two women. They were both experienced, capable, industrious, and trustworthy. When I would leave for home, they would still be as busy as ants with winter coming on. I knew the job would be done.

After two years as Control Foreman, I was moved back to PBX installation foreman. Those in power must have figured it was time to put me back into harness. Frank Appl had retired, and so I was given his crew.

Once more, Jay Worden reported to me. As always, he was not only a valuable employee but a good friend. Jay was teamed with Clarence Enright, another excellent technician.

One day they were working on a job that was both time consuming and difficult and so I sent John Dorsey by to give help. About mid-afternoon they completed this job and headed for another. Jay's sinuses were giving him fits and so he whipped into a drug store parking lot, planning to purchase a packet of Contact or some such over-the-counter capsule. John parked behind them and said, "Let's have a cup."

The others agreed.

Now the Company had rules on about every move an employee could make, including break time. If they had been on the job for more than an hour in the morning or an hour after lunch, then and only then was it permissible to stop for fifteen minutes, maximum.

Never, never, never, were employees to gang up. Any time two trucks converged on one coffee shop, that was considered to be ganging up.

By this time, Ma, in Oklahoma, had split between business and residence. The two Division heads took turns driving about, checking for company vans that were improperly parked or folks ganging up at coffee shops. This was the residence man's day. He spotted two trucks pulling into a coffee shop parking lot and decided to time them.

According to his timepiece, some sixteen or seventeen minutes ticked away before the three returned to their trucks. By then, the man had vehicle numbers, license numbers, and was ready to see that the violators were punished or at least reported. Notebook in hand, he approached them and said, "What are your names?"

"Who are you?" asked Jay.

This lack of recognition must have hurt a bit, but the man remained calm and told them his name and the extent of their transgressions. They dutifully provided their names and drove away.

When they came in that evening, Jay told me what had happened. I figured trouble was a brewing. It was.

The next morning my boss, Joe Floyd, told me that his boss, Kathryn, (A woman district head yet!), wanted me to suspend these three fine employees for one day without pay.

I argued; he said it had to be done.

When I refused he mentioned insubordination.

I asked for an audience with Kathryn; he said to call and see if she had time.

She had time. A half hour later I was in her office, pleading my installer's case. Of course by now they were out working and so it was too late to play the suspension game that day.

Unfortunately, there was always tomorrow.

I told her that these were good employees and that suspending them would do much more damage to morale than what good it would accomplish.

She said they had been spotted by a Division head who had informed his business counterpart and her boss had passed this gem of information on to her, expecting action to be taken. Now how would it look if she was to do nothing at all?

We argued, discussed, pleaded, threatened, for well over an hour. Finally Kathryn agreed that if I would first talk to each of these wayward employees, warning them that if it every happened again the consequences would be dire, and then document the conversations and see that this was placed in each employee's personnel file, she would let them off.

Kathryn was not happy with this conclusion, nor was she pleased with me. Oh well, by this time I had all but given up any idea of ever becoming Vice-President and General Manager for Ma in Oklahoma.

SIXTEEN

THIS 'N THAT

In the final days of her glory, Ma Bell had some million children fighting for their share of the sugar tit, often accusing siblings of getting more than their quota of Mother's money and affection. In spite of squabbles and feuds, when the chips were down, we would bust our collective butts to provide service. Even so, to many a customer there must have been times when it seemed as though we were not only spinning our of control but plotting against 'em.

Whenever Company policy and customer service clashed, something had to give. There were times when we would bend rules a little to soothe ruffled feathers; other times we couldn't, or wouldn't. Working in attics is one rule that few of us would ever bend.

There were several reasons why we wouldn't work in those not floored, lighted, and ventilated. First, there was the safety of the employee to consider. Also, if some big-foot telephone installer stepped through the ceiling, Ma would be liable for expensive repairs.

A primary reason though is that most attics are filled with blown insulation. Should a sweat-soaked workman crawl through this rock wool or fiberglass jungle, he would in all probability come out smelling like a man without deodorant. The next customer would then cringe at the thought of letting this weird individual with rock wool eyebrows into her home.

While working as an installation foreman, late one evening a Business Office Rep phoned informing me that we had a hostile customer. The man had ordered an extension phone in his home. To get where he wanted to go, the wire would have to be pulled through the attic or stapled on the outside of the house.

This was back before cordless phones.

The installer had surveyed the job and refused to crawl the attic. Would I go out and pacify the poor man? Yes ma'am. At least I would try.

It was a sprawling two story house located in Nichols Hills, exuding all the warmth of a mausoleum. Our customer was a vice-president for an oil company large enough to have a listing on the New York Stock Exchange. He figured that he pulled enough weight to get what he wanted and surely he could intimidate one pitiful little installation supervisor. I explained Company policy.

After I had explained, he exploded. "I will not have wire run on the side of my house as though it were some shanty in nigger town! Do you understand me?"

Hmmm. The man was a racist too.

"Yes sir. If you like, we will leave the wire and you can pull it in or have someone else do so."

"No! No! I'm paying your Company to install my phone. That means you pull the wire."

"Not in an attic, we won't."

"You *will* pull the wire through my attic!"

This time I did not lose my temper. "Good afternoon, sir," I said, and took my leave.

As I was walking towards my car, he shouted, "I'll call John Parsons. You'll be back!"

John Parsons was the Vice-President and General Manager over all of Southwestern Bell Telephone in the State of Oklahoma. Quite naturally he hobnobbed with vice-presidents and even presidents of prominent oil companies.

I doubt that I had driven much farther than around the corner before this demanding customer was on his phone, dialing Mr. Parson's number. Soon after returning to my office, the phone rang.

It was Mr. Parsons. Sometimes he preferred going straight to the person involved in a complaint instead of wading through cumbersome lines of organization. I also preferred doing business that way. He asked what had happened.

I explained the situation. "Good enough," he said. "That's what I needed to know."

My respect for Mr. Parsons rose considerably following this conversation. What he told our bellicose customer, I do not know. John might travel in the same social circle as did this egotist, but I doubt that their relationship was even close to being buddy-buddy. It's the last I ever heard on the matter.

Yeah, we had family feuds now and then. We had to walk a tight rope. Don't antagonize those good folks in other departments or you'll never get their help; stay after them to get all you can get. Too often we pulled in opposite directions.

As V. B. Sweatt was fond of saying, "Somehow or other we make money in spite of ourselves."

Maybe it's all in how we sold ourselves to the public.

Ma's public relations folks were always after some catchy copy in an attempt to catch the attention of a potential buyer. Sometimes it backfired.

One in particular was a thorn in our side. It was, "We may be the only phone company in town, but we try not to act like it." Sure enough, any time company policy came into conflict with customer wants, we would hear, "Yeah, you're the only phone company in town, and you darn sure act like it!"

Nancy and Judy, two ladies in the Business Office, wrote all service orders for small business in our district. Redheaded Judy usually remained calm. Nancy didn't have red hair, but she had the fiery temper.

When I was an installation foreman, Nancy phoned to request a short interval for a customer. I told her that I would not help, that this customer would have to wait his turn. After arguing for a few moments, she shouted, "Mr. Bross, you can go to hell!"

And then she slammed down the phone.

Well, now! Hardly the way to win friends and influence installation foreman. Still, as I considered the situation, I realized that Nancy must be under extreme pressure to do a favor for this customer. Without having any good reason to do so, I had refused to help.

So, I called Nancy back and told her to go ahead and expedite the service order. Neither of us apologized to the other, but from that moment I began to recognize the pressure these ladies were under and started being more cooperative.

Most first line supervisors hoped to take one more step up the organizational ladder. During my twenty years in management, there were numerous times when I was told by my immediate boss that I was on the list for promotion. Invariably there would be some reason why it did not happen. More often than not, the job would go to a Jet. And then we had to contend with "Affirmative Action".

"Affirmative Action", an agreement reached between Uncle Sam and Ma Bell to guarantee that women and blacks and Mexicans and other minorities or those who were only called minorities (eventually this category included nearly everyone living within the United States other than white males) filled a numerical quota, didn't help the chances of a white male hoping to move on. Suddenly there was no demand for us. Because of our sex and color, we were discriminated against.

Still, my bosses dangled the carrot in front of the donkey. Okay, I had the mentality of a jackass. Otherwise I would not have believed their propaganda. Until my last two years with the Bell Company, I never gave up hope. By then, it was too late.

Ollie Taylor would have made a good second line. So would have Frank Appl and V. B. Sweatt and a few others who come to mind. In my opinion, they should have been promoted. Like many others, they got caught in the squeeze.

When a man (or woman) was promoted to that second level, he or she had to be brainwashed. Upon completion of their indoctrination, they would repeat banal sayings that had, no doubt, been a part of their final exam.

Two jewels often heard were: "I'm not out to win a popularity contest; I just want results." Or: "I don't get ulcers; I give 'em."

When parroting the first, a new convert seemed to believe that it was a mortal sin to be liked even a little by his subordinates. During their first few weeks as a second-line supervisor, some would try extra hard to prove that they were meaner than a rabid dog. Given time, they would return to their normal self.

Usually.

As for giving ulcers, I questioned the wisdom of this statement. It was one of Harvey Faught's favorites. Of course it was only a figure of speech but I couldn't pass up the opportunity to take a dig.

One day we were riding together and after he had said for the hundredth time or so, "I don't get ulcers; I give 'em," I said, "Harvey, I don't see what we gain if we give ulcers to others."

"Whata ya mean?"

"Even if you don't feel compassion, it's one heck of an expense for the Company. Man takes off sick, we not only lose him for the duration but the Company pays his medical expenses. To me, this doesn't sound like a good business practice."

Harvey shot me an impatient look. No doubt but what he inserted a note in my personnel file, stating that I was unfit to even be considered for that next step up the ladder.

Harvey had this hang-up about vacations. He never took all that he was entitled to take and hinted that it would be nice if his subordinates would follow his fine example. One day he bragged to me that the year before he had given two weeks to the company. I told him he was crazy.

Offended, he said, "Why do you say that?"

"Did your boss take all of his vacation last year?"

"Well, yeah, but... "

"Did his boss take all of his?"

"I suppose. Well, yeah, he did."

"How do you think they feel about you not taking all the vacation you were entitled to take?"

"I don't know."

"They're laughing at you, that's what. They think you're a sucker. You've earned what vacation you get by sticking it out this long."

Harvey was silent for a moment and then he said, "I believe I'll take all of my vacation this year."

He did. Ma survived. Unfortunately, several years later, old Harvey died. Ma even survived that.

Over the years Ma has been attacked often from without and occasionally from within. When the attack came from outside, Ma was able to defend herself. When her own children turned against her, this was a different story. How do you discipline a recalcitrant child without making yourself look mean spirited?

In 1976, a book bearing the title *The Phone Book*, claiming to be a true story, was published. Having heard of the work and being the curious sort, I looked for and found a copy. After reading it through I decided this book belonged in the fiction category. Yes, it did! Supposedly this account of Ma's wicked ways was written by a man who had once been a child of Mother Bell's.

I say "supposedly", since the timid author hid behind the nom de plume of J. Edward Hyde.

Probably a little play on words here, as he tried to associate himself with J. Edgar Hoover and thus portray himself to the world as a famous investigator.

If so, it didn't work.

In his book Mr. Hyde states that telephone jobs are boring. The only bored people I have seen are those who try not to do any work. Thankfully, they were few. Mr. Hyde must have landed right smack in a nest of these folks. Perhaps he set the example.

He talks repeatedly of goof-ups in service and interpretations of Company policies. When a company has a million employees and handles a multitude of customers daily, sure, there are going to be foul-ups.

He implies that many promotions are given to bed-hoppers. If this happened, it was rare indeed.

He makes an issue of Ma's children conducting their personal business while on company time. In my experience, it was quite rare for any non-management person to do so. It was all but impossible for a supervisor, especially one who handled the outside folks on a daily basis, to draw the line between his time and that of the Company.

There have been abuses. Of course there have. However, the practice wasn't nearly as prevalent as he would have you believe.

Mr. Hyde! Shame on you! You bit the hand that fed you. No way to treat your ma.

Mr. Hyde (or whatever his name) claims to have once been a middle management employee. This is an ambiguous classification. Some second-lines think they fit into this category. Whatever his level, I think he gives substance to the theory that too many of us are often promoted to our level of incompetency.

From whom is Mr. Hyde hiding? Although he is openly critical of others for conducting outside activities on Company time, he boasts of having spent three years on the job gathering a "wealth of data" to use against the

Company. Supposedly, he was then fired for refusing to return his collection. I wonder if he returned the wages he didn't earn during those three years.

Bet he didn't.

It's doubtful that this book of his ever reached the critical acclaim or success Mr. Hyde had hoped for. As Ma's children, we might criticize some of her actions to one another, and we too often did, but it was taboo to expose your dirty laundry to the public.

At one time there were four installation foreman working out of the work center at 2929 N. W. 39th. Besides Frank Appl and myself, there were Joe Carey and Roger Horton.

One morning after our crews all hit the road, we decided to go to a restaurant on May Avenue. Frank and I only ordered coffee while Joe and Roger decided to eat breakfast. After their food had been set before them, Joe began eating his pancakes while Roger cut into his Western Omelet. Roger had taken two or three bites when all of a sudden he became quiet, had a sick look on his face and laid his fork aside. He was looking right into his omelet.

There for all to see was a Band-Aid, probably one that had fallen from the cook's finger, since it was round and of finger size and was still stuck together. After the waitress had been summoned, she called the manager. They both apologized and offered to replace Roger's breakfast. Roger didn't want another omelet; Joe didn't even want another bite of his pancakes.

After that whenever we went into a restaurant, someone would invariably say, "Bring Roger a Western Omelet and hold the Band-Aid!"

No matter what their party affiliation, we were never anxious to see bigwig politicians, such as the President or Vice-President of the United States, stay overnight in our city. They don't just drive up and check into a hotel or motel the way us common folks do. Their impending visit always requires a great deal of advance planning and the installation of a communication system that is complex and takes time to install.

Prior to one visit in 1980 by a Vice-President who was then reaching for the grand prize, we were installing telephones in luxurious suites for him and his aides. The head advance man had yet to arrive.

Whenever his name was spoken, those who had come to pave the way would tremble. No doubt but what his subordinates lived in fear of the man. His suite was located near that of his boss.

Jay was working in the man's suite when he made his noisy entrance. His first words to Jay were, "Who the hell are you?" Jay straightened up and said, "Who the heck are you?"

The man made some obnoxious remark and Jay said, "Hold it a minute. I've been hearing all day that you're a regular horse's rear. That's all right with me, because I'm a horse's rear too."

The man frowned for a moment and then broke into a smile. "Is that what they say about me?"

"That's what they say."

He stepped forward and shook Jay's hand. "Good, that's the impression I want to leave."

Shoot, we got along just fine with the old boy.

While I was an installation foreman, I began wearing slip-on shoes. Being a country boy, often, when I sat down, off they would come. Usually I wasn't conscious of having sock-clad feet. There were times when this was inappropriate.

One morning, Ollie, Kubiak and I went to a fancy cafe for a cup of coffee. The place was packed. Kubiak and I were discussing something or other when Ollie slipped off to the restroom. When it came time to leave, my shoes were missing. There sat Ollie, grinning a somewhat devilish grin.

"Where are my shoes?"

The grin widened. "Judas, can't you keep up with your shoes?"

"Dang you, Ollie."

I had to walk across the room in stocking feet to retrieve my shoes. He had placed them under a far table. It's not dignified for one of Ma's supervisors, dressed in coat and tie, to parade around in this fashion. Some folks thought it was funny

Another time I was sitting in my office, no shoes on my feet. Ollie grabbed them and was out the door before I could react. Even as I ran screaming after him, he tossed them both onto the roof.

I had to pull the ladder off of a truck and climb up after them. From then on, anytime Ollie was near the shoes stayed on.

When I flew in from New York for the weekend we made a scheduled landing at St. Louis. There was a short delay while a couple of Airline mechanics fretted over the hydraulic system. Seems as though they didn't fret enough.

After we had taken off, an announcement was made that we would land in Dallas, not Oklahoma City. The crew could not determine whether or not the landing gear was operating and Dallas was better equipped to handle emergencies. As we approached the lighted runway, we were coached on the crash position. Emergency vehicles stood by. Fortunately, the landing gear worked and we touched down smooth as could be. They shipped us to Oklahoma City on the next available flight.

My wife and mother were waiting at the Oklahoma City Terminal. Ollie was with them. When the airport announced that the flight had been diverted to Dallas for mechanical reasons, wife and mamma knew for certain that their beloved husband and son would crash into the Red River or some such place. Ollie kept them loose. As reported to me later, he said, "That ding-bat probably went to the restroom and pulled the wrong chain."

While I was working as a PBX installation foreman, Oklahoma City's business community was leaping into the future. As companies grew, they often changed to large, complex telephone systems. Many of these had to be switched to the new equipment after closing time. Consequently, we would often work late. One night after we'd switched to a new system I arrived home after ten. I expressed my frustrations to Dixie, even going so far as to call Ma Bell a "Heartless bitch."

Dixie was family, so saying this was permissible. She very patiently said, "She's provided us with a good living for a long time."

She was right. I had no reason to complain. Still, a man has to let off steam now and then.

The new car for Wewoka arrived not long after I'd left town. Frank Appl, the same man who helped me so when I was promoted to PBX Foreman was promoted into management and followed me to Wewoka and so he received the car. Late one evening the quality control supervisor, Harvey Pringle, called Frank and wanted to meet in Chandler the following morning to check some jobs. Frank decided to drive up that night so he would be there bright and early the next morning.

Darkness had fallen before he entered the little town of Cromwell. On his right he saw a horse and was watching to see that it didn't jump out in front of him. Coming toward him in the other lane was a car leading another horse. Horses everywhere. What Frank didn't see was the horse behind the one being led, not until they'd had a head-on collision, wrecking the car and killing the horse. Actually, it was a mare that was killed. The baby in her womb popped out and was killed too

Frank's new car had 54 miles on the speedometer. Although it is difficult to see how this accident could have been avoided, Comer had to bring the incident up at every supervisor meeting from then on, and continued to do so as long as Frank remained in his organization.

Joe Floyd is an okay guy who came up through the ranks and realized that at the second-line level he had reached his potential. He's a man who has a lot of common sense. Some of the time while I was a PBX Installation Foreman, I was working for Joe.

Soon after Joe took charge, it was nearing time for the State Fair. This event called for installing various telephones and radio circuits and such. Two of my hot-shot installers handled the installation.

Now that the Fair was underway they remained on campus to work additional orders and to handle repair as needed. While they wandered the Midway, some folks on staff who were responsible for safety in our Division decided to pay me a visit.

Staff people are supervisors without anyone to supervise. It is often difficult to define exactly what some of these folks do. Whatever it is, we felt as though some three-fourths of it we could do without. Some are quite competent while others seem to have little or no idea of what is happening.

Joe got word of their coming and so hurried over to offer moral support. The staffers arrived, one second-line and one who was District level. Neither seemed to know about the outside world of telephone workers and neither was capable of handling a crew and getting the job done.

They were as picky as could be and came up with some silly ideas that even a three year old child would discard as being absurd. Joe soon had stood just about all that he could stand.

Eventually, he said, "Don, I sure hate to run off and leave these guys, but me and you need to keep a customer appointment."

What appointment? "That's right, Joe. We sure do."

The two staffers just couldn't believe that we would leave them in the midst of such an important undertaking. As I drove towards the Fairgrounds, Joe said, "If those guys ever woke up in the real world, it would scare them to death."

Yeah, Joe's all right.

Gene Griffin was promoted to third-line and so he then replaced Kathryn as our District head. It was soon apparent that Gene meant business.

Our District wasn't doing too well on indexes. On several, we were dead last in the Division. That would soon change. First thing Gene did was to call all of his supervisors to a meeting at the Trade Winds Motel. We were fed lunch and then he took the floor.

Gene is a take-charge type person. There was no doubt but what he had no intention of remaining at the bottom of the totem pole. He made it clear that anyone who wasn't willing to make whatever sacrifice needed to move his District from its dismal position to the top of the heap had better be looking for another job.

When the meeting ended, everybody knew where Gene stood.

Not long after this, it was decided that Business Installation needed a control foreman. Joe informed me that I was Gene's choice. This suited me fine.

Gene gave me a pep talk and made it clear that he would not tolerate any missed due date orders. Even though I would remain a first-line supervisor, I would have the authority to dispatch installers as necessary, no matter what a particular installation foreman might think of my doing so.

For the first time since going into management, I had nobody reporting to me. It was too good to last.

It didn't last long. Eventually I was given added responsibilities. For me there would be no more moves. This is where I would stay until the end.

It wasn't far off. But first let's hear from other voices.

SEVENTEEN

OTHER VOICES

Having never been a part of Traffic, those faithful folks who routed calls on the *information highway* of the day, I am extremely grateful to those operators who offered to share their experiences. I am also grateful to those from the other departments who provide us with amusing or thoughtful moments.

Our telephone operators of times past played an important role in the development of the Bell System, a role that is often overlooked. Working conditions were less than ideal, but these ladies became known as *the voice with a smile*. They were together in a relatively compact space, often part of a crowd, and yet in a way they were somewhat isolated and alone.

In these days of air-conditioned cars, shopping centers, homes, and place of business, we tend to forget that this comfort is relatively new. An operator's work space could usually be kept warm in the winter but they could count on being cooking hot during the summer.

The earliest account I have from anyone from any department is from Vera Swanson, a friend from Guthrie, who started work for the Telephone Company in 1920. Vera says, "In the early 1920's, we didn't have any air conditioning and it would get very hot in the room where we worked as operators. One day the thermometer in our room said that it was 106 degrees. All we had was fans in cages and wet towels that hung over them. This didn't help much."

Of another incident, Vera says, "We had a rural board. The operator working that board took calls from and made calls to rural customers. My board was next to hers. One day there was a young lady working that board and for some reason she fainted. I kept her on her feet until the Chief Operator came to help."

Another good friend is Matilda Rodman, who began work at the beginning of The Great Depression. She says, "After graduating from high school in 1928, I went to work for Kress (a five and dime store) on September 15th of the same year. There we worked nine hours a day, six days a week for a salary of $8.00. Then one day Irene Wille, the Chief Operator, asked me if I wanted to go to work for the Telephone Company. It was January, 1930, when I made the change. There, for a five day work week, I was paid $10.00 and a dollar more when I worked evenings."

In Guthrie, the operators worked upstairs at 111 South Division. Although the room had several windows, the operator's positions covered many of these. Matilda says, "That was always the hottest place. We had some overhead fans, but all they did was stir the hot air. In the summer, I would always come out of there at the end of the day feeling all washed out."

Matilda predated the Union, the Communication Workers of America. She says, "I was Chairman of the Association when they joined the Union. I resigned from the Company in 1943 and was always glad I didn't have to go on strike."

Matilda's reason for resigning was one that was more common then than it is today. She became a mother and wanted to stay home and raise her son, a move that she has never regretted.

My cousin, Vickie Douglas, was an operator in Guthrie for the better part of ten years, beginning in 1941. Vickie says, "What I remember the most is the day President Roosevelt died in 1945. I was working the board and it lit up as it had never lit up before. It looked like Christmas tree lights."

Across the street from Guthrie's Telephone Office was the White Rock Cafe. Vicki says, "Some of us would call over in the morning and order breakfast. At the beginning of our fifteen minute break we would rush over, jaywalk over, and breakfast would be waiting. We had to hurry and eat so we could get back in time. We didn't dare be late."

Ludmila Hrdina (Tinkerbell) started work for the Bell System On March 20, 1941, some eight and a half months before the beginning of World War II. After the war started she was getting so many calls from Tinker Air Force Base that they began calling her 'Tinkerbell'.

Before going to work for Ma Bell, Ludmila was doing domestic work for $5 a week. A starting salary of considerably more from Bell was certainly welcome.

Ludmila was at the board when World War II broke out. Referring to her switchboard when the news first hit the mainland, she said, "Lights came in, looking like lightning bugs on the farm on a dark night."

Ludmila worked until 1975, all but ten of those years in Oklahoma City as an operator. Those ten years were spent in Sacramento, beginning in 1951. Ludmila says, "I was working in Sacramento when we started using area codes. We were trained to ask the customer if they had the area code and encourage them to dial the number themselves. One man called in and asked for a number in another town. I asked if he had a code. He said, 'No, I don't have a cold. How are you today?'"

There was a time when mobile phone calls came in on Ludmila's position. Another operator transferred a customer to Ludmila because this operator thought the customer wanted the mobile phone department. Turned out the customer wanted Mobile, Alabama.

Betty (Morrison) Enox worked as an operator at Durant. She says, "It was New Years Eve and only two of us were on duty as midnight approached. We knew that when the hour came the boards would light up as everyone would want to call someone and wish them a Happy New Year. This time the customers had to wait, for at the magic hour my coworker and I removed our headsets, hugged and wished one another a Happy New year."

Betty tells of a time when a man reached her from a pay phone. "Operators were *never* to complete a call unless the customer deposited his nickel," she said. "This time a customer came in on my position and the man was so excited he could hardly talk. Finally, I understood, 'I don't have a nickel and I have to make a call. There's a tornado on the way.' I let him make the call and sure enough a tornado hit soon thereafter."

Margene Baker started work as an operator in Edmond soon after the new office went in at 14 East First. There were thirteen outward positions and one inward. This inward position provided access to Edmond for all operators outside of town, including Oklahoma City. "Yes," Margene says, "it was a long distance call to and from Oklahoma City. No one had ever heard of 'wide area'."

In those days they trained operators the same way they trained linemen. No formal training. She was trained by more experienced operators, one of whom hated to instruct newcomers. If Margene would mishandle a cord, this operator would hit Margene's fingers with a ruler.

Patience was called for. An operator would plug into a lighted jack and say, "Number, please?" If the customer didn't know the number, they would give a name, then the operator would look it up and ring through.

Since there were a half dozen multiple positions, an operator, before plugging in, would test the jack by touching the edge of the hole and getting a little buzz. Margene says, "Occasionally we would be extremely busy and would plug over the signal and then the person we had on the line would hear the voices of the people talking. Sometimes that was bad. Edmond was a little 'Peyton Place' as probably every other town the size of Edmond was. As we learned the town, there was no way that affairs could be kept secret from us for if people used the phone at all, we knew who had called whom. We surely learned to keep our mouths closed tightly."

Of another incident, Margene says, "When pay phone customers called long distance, we had to notify them when they had talked three minutes because they paid for the initial time. The phrase was, 'Your three minutes are up, please signal when through.' One operator went in to notify a customer and her mind went blank. She said, 'Your time has come.' The customer hung up as usual but we never did know what he thought had happened."

This, too, was in the days before work space was air-conditioned. Margene says, "It was very hot in the summertime. For an air-conditioner, they reluctantly let us have a wash tub of ice with a fan behind it. The ice melting was supposed to cause dampness and it was hard on the cords, they said. It was so hot, without even an evaporative cooler. We could not wear shorts; sleeveless dresses were frowned upon. That's why I hated summer."

Concerning the earthquake that hit the area in the 1950's, she says, "I was working that morning, standing rather than sitting at the high stools at our positions. I felt this strange sway and I was a bit frightened. I had never had that feeling before. I remember wondering if this might be the Second coming of the Lord. But in less than sixty seconds, I knew that if it were everyone in Edmond was taking time to say 'Good-bye" because they had all picked up the phone."

Margene says, "In 1961, Edmond cut to dial. An article in the paper said that Edmond would not miss 35 operators. We were all angry at the editor because we felt that Edmond would miss us."

Margene, I would bet that Edmond did miss you all.

Another operator with interesting experiences is Reba Brown. She spent part of her career in Traffic and the remainder in the Plant Department so she knows the problems and joys experienced by those in both places.

33 of Reba's 36 years spent with Ma Bell were years with perfect attendance. Not many of us could match this.

She recalls a call she handled to an individual who was supposed to be at an Ardmore Hotel. We'll call the man, John Wilks.

Reba says, "This was a person-to-person call and so I rang the hotel and asked for Mr. Wilks. The clerk who answered said, 'Mr. and Mrs. Wilks have already checked out.' Before I could use my split key, I heard the lady making the call say, 'I'll have you know *I'm* Mrs. John Wilks.' By then, it was too late to split them off."

When Reba retired, the following was written for her and she has allowed us to use it here. In its entirety is:

NOT FUNNY AT THE TIME

In 1950 Edmond was a small town and it needed only a small office and only one operator to man the office in the wee, wee hours of the morning. Reba was that operator. She worked the 10 P.M. to 6 A.M. shift.

This particular morning, Reba's relief called to tell her she would be late. It was near six and Reba knew her husband, Corky, would be outside waiting. So she ran to the door, propped it open with her toe, stretched out as far as she possibly could and yelled at Corky, "I'm going to be late!" Corky yelled back, "Huh?" "I'm going to be late!" Corky still could not understand what Reba was trying to tell him. So Reba stretched out that last little millimeter and yelled, "I'm going to be...!" Click! Guess what? The door closed. I mean I'm talkin' CLOSED, SHUT TIGHT AND LOCKED!

Well, if you know Reba, you know she panicked with a capital "P". Corky found her standing there sobbing her heart and tears out. She cried out, "Oh, Corky, I can't get back in! What am I going to do? They're going to fire me, I know it! They're going to fire me!"

Remember now and understand this: At 6:00 in the morning Reba was the only operator in town. There was no way anyone in Edmond could make or receive a phone call without going through Reba. But now, even Reba couldn't make or receive a phone call because she wasn't at her position which was on the other side of that CLOSED, SHUT TIGHT, LOCKED door. GRIM FACT: The whole town of Edmond was without phone service.

Corky was finally able to calm Reba enough to get her to understand that the only thing they could do was to go see the Chief Operator. UN OH! DOM DE DOM DOM!!! (We'll leave this to your imagination.)

Well, it all worked out for Reba. Needless to say, she didn't get fired. To everyone's amazement, no one called to report not having phone service that morning.

"Way to go, Reba!"

In 1973, that same Bill Kincaid with whom I'd worked and ridden with in the back of a construction truck, was third-level management and in charge of the coin department. Midge Sullivan says that Bill offered her a management job; if she accepted she would be the first woman manager in the coin office for all of Oklahoma. Bill offered a few words of caution before she decided.

He reminded Midge that her coin collectors would consist of one lady and fourteen men. If she were riding with any of these, she would see a side of life she was not accustomed to seeing. If they were out on the road and had to stay overnight, she might find that some of the men reporting to her would lead a night life of their own. They might party and go into places she didn't even know existed. Midge told Bill she was sure she could handle the situation and wanted to give it a try.

Midge says, "The first time I rode with a man coin collector, he was working the Fort Sill route. We would stop, collect the money from a phone, stick in a new box and then go on. Some two hours into the route, I called my office and was told that one of our customers had phoned the area office and said, "I don't want to get the young man in trouble who was here collecting my pay phone, but you should know he had a woman riding with him.' We all got a laugh out of that."

Now that Midge was in management, one of her jobs consisted of "salting" the pay phones, putting coins that had been chemically treated into the upper housing to see to it that our workers in the field were honest. On this particular day, she was working in the Lawton, Fort Sill area.

Midge says, "I couldn't go out in a Company car or some of the repairmen might see it, recognize me and know what was going on. I rented a car and went into some very seedy places, including bars and tattoo parlors and houses of prostitution. After I'd made a number of stops, a Lawton policemen pulled me over. He had been observing and wondered why I had been visiting these places. I explained what I was doing and why I was not in a Company car. He went to his car and radioed in and the next I knew a lady policeman came out. The two of them questioned me and it soon became clear that they thought I was a "Madam" out visiting her houses. Finally they called the office in Oklahoma City and decided that I was who I said."

Midge says an even bigger surprise was when an Oklahoma City Repairman was working on a pay phone on North May in Oklahoma City. He came into the office and told Midge that the key he had wouldn't work. She went with him to try her hand.

Midge says, "I thought perhaps I could pick the lock if that's what was needed. We went into the bar and I was ready to put the key into the lock when the music started and the repairman who had brought me, said, 'Look.' I turned and there on the stage was a tall, skinny girl without a stitch on. I hadn't known this was a strip joint. When I put the key in the lock, it opened right up. I don't know if the repairman deliberately got me into this place or if he was so excited at what he was seeing that he couldn't get the key to work."

Bill Day is a man of many talents who has done everything from lineman to construction foreman to PBX Installer. It was while he was a construction foreman in late 1948 that this amusing incident took place.

Bill says, "We were working up at Enid on an ice break and had been away from home for a while. About six of the wives got together and decided they would come up on a bus to see their husbands. Frankie, my wife, didn't think she would be able to come because we had a baby, who is also named Bill.

"Anyway, she came to Enid and brought the baby. Next came the problem of what to do with little Bill. They looked all over the hotel and couldn't find a crib or baby bed of any kind. We thought about putting him in a suitcase but somebody said there was always the danger the lid would fall shut and smother him. So, we didn't know what to do.

"I got this idea from Moon Mullins in the funny papers. Moon's little brother always slept in a dresser drawer, so that's what we did: we left a dresser drawer partly open and fixed him a bed inside. It worked out just fine."

Martin Baucum recalls a couple of incidents, one he only heard about and the other was one he took part in. First, Martin tells the story he heard, about the three guys working together. Of the three, one was unusually tall and always wore a cowboy hat and boots, a second man was part Indian and the third was maybe a little shorter than average.

Martin says, "They were working together, I think they were trimming trees. They got through before time to quit and pulled into this alley to make out their time reports and to kill a little time. There was a progressive, kind of a stairstep hedge between the alley and a yard. The three of them started at the low end, jumping the hedge and working their way up.

Anyway, the lady called the Telephone Company and said, 'I want you to know, there's a damned cowboy, an Indian, and a midget out here and they're tearing up my hedge.'"

This next story Martin experienced first hand. He says, "Buril Hillis and me were installing a key system at Hodges Truck Line, near the Canadian River. They had a large shop building and Mr. Hodges, Sr. lived on a hill in a nice house up from the shop. We needed to install an extension at the house so they could take calls there at night. We worked a couple of days, getting everything all done and we saved the job at the house until the last day. We got up there one morning when it was rainy and cool. Rather miserable, working outdoors.

"Mrs. Hodges met us and said, 'Make yourselves at home. I'm sorry I have to leave, but I have an appointment. I would fix you something, but I've got it all laid out. Coffee pot's in there and coffee's where you can find it and there's cookies. When you get ready for a break, just come in and make yourself at home.' She felt sorry for us, having to work out in the cold.

"She left and we did get cold. When we got all the outside work done, we went into the house and hooked up the coffee pot and got the cookies out and we were sitting there, and I may of even had my feet propped up on a chair, or something. In came Mrs. Hodges. In she walked and said, 'What in the world are you guys doing? You come in and just take over somebody's house like it was your own! What gives you the right to do that?'

"We were speechless. Here was the lady, or so we thought, who had told us to make ourselves at home, to eat and drink and rest. Now that we were doing so, she was really getting onto us. I didn't know what to do. I would have crawled into the floor, if I could. Buril, too. We were startled. And then she started laughing. This wasn't Mrs. Hodges at all, but Mrs. Hodges' twin sister. They were identical. She knew why we were there and knew we'd been invited into the house. After laughing, she managed to say, 'I really got you, didn't I?' And she did. Did she ever!"

Tom Fowler, a former co-worker and fellow author who now lives in Texas, provides us with this amusing anecdote. Tom says, "From November, 1973 until

July, 1977, I worked for Howard Owens at the 1004 NW 79th work center as an installer. One morning a very funny thing happened.

"It was Monday morning. The previous Saturday, somebody had installed a telephone in a new house and laid a temporary drop wire on the ground. A note told Howard a return visit would be required to make a permanent connection. The installer wrote neither the subscriber's address nor the service order number on his note and identified himself only as 'LBS'.

"Having temporary drops on the ground was a serious matter, as Southwestern Bell had been the target of several lawsuits because of people tripping over them. So, it was no small matter to Howard when he could not determine the identity of the customer or of 'LBS'.

"Howard had been attempting, without success, since 6:15 to identify the mysterious 'LBS'. 'Probably someone from another district,' he lamented, as he asked if anyone knew.

"Larry smiled sheepishly. 'That was me, Howard,' he said. 'I thought you knew.'

"But Howard didn't know that Larry would sometimes sign his memos and notes as 'LBS'. We all had a good laugh at Howard's expense when someone suggested that from now on Larry sign his notes 16 ounces.

"Have I mentioned that Larry's full name is Larry Pounds?"

Going all the way back to my first days at Sayre, a lineman who was on the crew was Joe Guilliams. Years later, as our paths crossed, Joe would say, "There you were, wearing blue overalls, a ball cap and a pair of old work shoes. The cap was pulled down over your eyes and tight in back. You were the hickiest guy I'd ever seen and so country we had to pick hayseed out of your hair."

I suppose they did at that.

Vernon Shaw recalls a time when working for Smiley and we had been loaned out of town for a few months, first at Chickasha and then on to Lawton. While at Chickasha, one of our jobs was to add four copper toll-line wires to an existing toll line that ran alongside the highway from Chickasha to a little town by the name of Alex.

It was the coldest day of the winter but Smiley decided it wasn't too cold for his linemen. No sireee, when it came to working in inclement weather, he was willing to put his men up against those from any crew. Of course while we were freezing, he was sitting in the truck with the motor running and the heater on.

We built a fire alongside the road so that we could warm up now and then. Smiley managed to wake up enough to roll his window down and call Vernon over to inform him that we couldn't build fires along the road. Of course once he'd said this, the window went back up.

Another incident that is still vivid in Vernon's mind happened while we were with Smiley on the ice break to Wichita, Kansas during the winter of 1954 and '55. The weather was bitter cold and ice covered all the outdoors.

On this particular day conditions were so bad that every foreman other than one opted to keep his crews at the motel rather than take a chance on an accident. That one foreman was Smiley. Heck, the weather didn't get too bad for his boys. More than likely we were the only construction crew working in the entire State of Kansas that day.

Out to the country we went, the truck slipping and sliding on the roads. Ice was still falling. There was no traffic; other folks had more sense than to get out. When we reached the work site, Smiley told us what to do then climbed inside the truck and turned the heater on high.

We staggered from pole to pole, shivering and shaking. Vernon, after descending a pole, twisted his knee. Painfully. Severely. When he reported this accident to Smiley, he hardly endeared himself to the old man who was determined to take every means possible to not report an on-the-job-accident.

Vernon says, "He wouldn't take me to a doctor. I stayed in Wichita until we all went home but I had to stay at the motel since my knee was swollen and painful. When we got back to Oklahoma City, he still wouldn't take me for medical treatment. I hobbled around for a month or so grunting for you guys until finally I went to the Union and they saw to it that the Company got me to a doctor.

"The doctor said that if he'd had me the day after the knee was injured he could have made me as good as new. Because so much time had passed, he couldn't. I got a thirty percent disability out of it. A supervisor told me years later that I would never get a management job because of the disability."

If what this supervisor told Vernon was true, then it's a shame. Vernon had all the attributes to have made a good management person. And yet the Company kept old Smiley in a management position until the day he retired.

Hard to figure.

A person could get a headache just trying.

Vernon goes on to say, "Nobody now could ever understand how hard we had to work back then. We had to dig our pole holes by hand. One day we were near downtown and had to bust through a good four inches of concrete before reaching packed dirt. I was using that eight foot digging bar and even though I was wearing gloves, when I went home that night my hands were raw and bleeding. I'm sure yours and everybody else's were too."

I'm sure they were. Thanks, Vernon.

Dee recalls an incident that gave him a jolt at the time. It happened while we were working the ice break at Jonesboro, Arkansas. Dee says, "We were working alongside the railroad tracks and there was even water over the tracks. They didn't have enough hip boots for us all so you were wearing the boots and I was up on the tracks wearing overshoes. You waded over to the pole and about the time you got there you went under the water, completely out of sight. Just as I was ready to jump in and see if I could find you, up you came and right on up the pole. Somebody had left a hole open and you'd stepped off into it."

Dee has a lot of stories to tell. It's a pleasure to talk with him and recall those days of long ago. In lots of ways, they *were* the good old days.

Frank Appl recalls an event that happened in the early '50s. Frank was working north of Edmond as a lineman for Red Hammett. It was a windy day and the grass was dry. Prior to climbing a pole, Frank, who had been smoking a cigarette, flipped the butt. While he was up the pole, a grass fire started. The pole didn't burn but about everything else around did, including about a half mile of a farmer's fence.

Later, Frank told his boss, "I think I started that fire."

Red decided to make amends so the next morning they pulled into the farmer's yard with a trailer filled with used but good crossarms. To the farmer, Red said, "I think one of my men started the fire. We brought out some crossarms you can use as fence posts. Where do you want them?"

"I want them in the ground and the wire fastened to them," said the farmer.

Red's crew, including Frank, spent several days setting the crossarms as fence posts. Instead of stapling the barbed wire, they screwed insulators on and tied the wire to them. When they had finished, the farmer had one of the prettiest fences in the county.

Thanks to everyone for the anecdotes you have provided and to all of you who made this book possible. Now let's return to the death of a legend.

EIGHTEEN

THE END DRAWS NIGH

January 8, 1982. Folks throughout the world must have been surprised by the news, but we were stunned. Break up the Bell System? Absurd! Not possible! Like the Titanic, she was thought to be unsinkable.

Our vision was of a company who had maintained a low profile, preferring always to sail a steady course and thus seldom catching the attention of market speculators. Its stock was said to be safe for "widows and orphans" since it had paid a dividend since shortly after creation. And now it was going down the tube.

Her children felt betrayed. The divestiture of her operating companies, abandoned by Ma Bell like kids no longer loved, was reported to have been agreed upon in a consent decree. Unlike a judgment that is arbitrarily handed down by a court, in a decree all parties have to agree to the terms.

Did Ma surrender? Or did she have a death wish?

At the time of her demise, Charlie Brown was Chairman of the Board of AT&T. When he took office in 1979, some five years after Uncle filed suit, he made a statement that should have been a forewarning, had we but listened. He said, "Ma Bell is a symbol of the past. Mother Bell simply doesn't live here any more."

Ouch! He was only being realistic, but it still hurt. Perhaps she no longer lived at 195 Broadway, New York, New York, but the old girl was still well thought of in the state of Oklahoma. By most folks, that is.

This was not the first time the Government had declared war on AT&T. Before, we had always fought back. As early as 1913, Uncle had vowed to tear the heart out of Ma. In that year, the Kingsbury Commitment was entered into.

To save her soul and remain intact, Ma agreed to abandon her goal of monopolizing communication in the United States. Yeah, she even relinquished her control of Western Union.

In 1949, Uncle tried again. Ma was getting too big for her britches. Sam wanted her to divorce Western Electric, the subsidiary who manufactured nearly all of her material and equipment.

Seven years passed before another consent decree was reached. (Here we go--always a consent decree.) Ma could stay alive if she agreed to forgo the computer business. Since computers were then thought of as little more than a passing fad, this didn't seem to be much of a sacrifice.

Ma really showed her vulnerability in 1968. That's when the FCC returned the Carterphone decision. For the first time ever, it was legal for certain equipment not manufactured by Bell to be attached to our circuits. It wasn't long before the attacks became more numerous.

The following year a company with some two dozen employees, by the name of Microwave Communications, Incorporated, was granted the right by the Federal Communications Commission to enter into the intercity "private-line" business between Chicago and St. Louis.

Little did we know that this simple action set the stage for Ma Bell's eventual disintegration.

Now certain folks in government felt that Ma Bell should be brought down a notch or two. AT&T was the largest free enterprise operation in the world, and therefore made a tantalizing target. These folks encouraged this new company until it became a competitive force.

Then came a time when the U. S. Government was in total disarray (When isn't it?). Nixon had resigned and Ford had taken charge. In November, 1974, Uncle filed suit against old Ma, his objective being to strip her of her Operating Companies and anything else he could lay hands on.

Never mind that here was the most efficient phone company of any size in the world and possibly in the entire universe. Never mine that Uncle was so fat and lazy that he could hardly move, much less think straight. Like a compulsive eater, he kept adding to his enormous weight by padding payrolls and creating bureaucracies that bordered on being meaningless, albeit dangerous.

By golly, we have this gaggle of lawyers, and we can always get more. That's one thing we have plenty of in this country. Let's point them at AT&T and see what happens.

So these Justice Department experts, law degrees from prestigious Eastern Universities decorating their walls, liberal enough to want to destroy any big organization, other than the United States Government, took on Ma and her covey of lawyers.

George L. Saunders, Jr. was hired by Mother Bell to lead the defense. George was an intelligent man who reportedly could speak in court for hours without benefit of notes or other props. Besides, it is said he sincerely believed in his cause and had little doubt but what he could win the legal battle and preserve the Bell System.

In February of 1975, the case went before Judge Waddy in Washington, DC. From all accounts, Judge Waddy cared not a whit who won, only that all legal procedures were followed and that justice should prevail. The lawyers for AT&T felt as though they would get a fair shake in his courtroom.

Judge Waddy, a black man who had worked his way up the judicial ladder, is said to have had very little patience with the tactics of the government lawyers. This was good for our side. But then an early question raised by Mr. Saunders was whether or not the U. S. Justice Department even had jurisdiction.

This wrinkle caught everybody off guard. Judge Waddy called a recess. It took our speedy Justice System only three years to answer that one question.

Heck, yeah, they had jurisdiction.

The situation was getting dadgummed serious. At the time, John DeButts was Chairman of the Board at AT&T. It seems as though he was the last Chairman we had who felt any affection towards old Ma Bell.

John decided it was time to make his voice heard in Congress, the Administration, and hopefully by the user public. His was a voice crying in the wilderness. Nobody seemed to care that our government was intent on bringing the Bell System to its knees.

Everybody just reached for their telephone and kept right on talking. Old John must have sounded like an Old Testament Prophet as he forecast an era of gloom and doom. Who is he trying to kid? AT&T is a hundred years old and invincible. No matter what the government does, they will survive.

By the time the case was reactivated in 1978, Judge Waddy was dead of cancer. Judge Harold Greene took the case into his jurisdiction. The Government's bill for prosecuting AT&T ran into the hundreds of millions of dollars. What the heck, it's only money. Let's print more and tear down this empire.

No help would come from Congress or from the Carter Camp. It appeared as though the case was moving toward a disastrous conclusion for AT&T when a wondrous event happened. Ronald Reagan was elected President. The only question now was how quickly would the Movie-Star President take action to dismiss the suit.

Sadly for us, Ma's children who stood helplessly by, his course of action was to take no action at all. Now that he was safely ensconced in the Oval Office, Ronald had to attend to more important matters. Although the President's advisors were divided on whether or not to call off the hounds, he heeded the advice of those who felt he should do nothing. Some folks feel he was quite good at this.

Back in the courtroom, monumental decisions were being made. It would seem that AT&T capitulated. They offered the Government the heads of the twenty-two Operating Companies if AT&T could still control Long-Lines, Western Electric, Bell Labs, and all that. And, oh yes, he would like to have a go at the computer market. Will this be acceptable?

The government could not believe how effective their gaggle of lawyers had been. They jumped at the offer. At the moment, nearly everyone was happy.

Nearly everyone.

From all reports, George Saunders was not a happy man. It seems as though he had been kept in the dark about these behind the scene negotiations and had continued working on his case in court. When he heard the news, it is said that George simply could not believe that AT&T and its Board of Directors would do this to him or to their own Company.

Before the press got wind of this historic event, the Presidents of all the Bell Operating Companies were summoned to New York where they were told what to many of us would become the doomsday story. Those men who would head the new Companies must have been quite pleased. Those who would not felt deceived.

It was not my lot to head one of these new Companies. I wasn't one to dance on the old girl's grave.

When an announcement was made public, soothing assurances came from both the Government and AT&T. Competition is always good, said the government.

"Amen!" said AT&T.

Local telephone rates will not increase as a result of this action said spokespersons for AT&T. Employees will not be adversely affected said spin doctors for both sides. Everyone will benefit.

Hogwash! Balderdash! And so on.

We children of a dying Ma were fed these lines, words that were intended to pacify before we discovered the truth. They truly spoke with forked tongues.

Gradually the puzzle pieces fell into place. As the days passed, many of these pieces constantly shifted. That first year, as we awaited divestiture, uncertainty gnawed at us. The second year was only slightly better.

The Presidents of the new Baby Bells, as these Companies would soon be known, had seen right off that without their usual subsidies from AT&T rate increases for basic service was a must. When it dawned on the general public that their relatively low telephone rates were going up, a great wail went up throughout the land.

The cries were so loud that even Congress soon awoke and discovered that while they'd been napping someone had destroyed the most efficient telephone system upon this earth and possibly the most efficient in the universe.

What's this? Who's responsible?

Congressmen jumped to the rescue. Even as Ma's structure crumbled, some tried to prop her up. What we need is committees to investigate, so these gatherings of inept congregations mushroomed throughout Congress.

After filling the halls with enough hot air to cut down on heating costs for a year or two, they pointed wagging fingers in all directions other than back at themselves. Their puny efforts were comparable to trying to stop an avalanche with a snow shovel.

It's possible that President Reagan said something along the lines of, "Well, hmmm!"

Meanwhile, back at the funny farm, we were running hither and thither, searching to see where pieces would fit into the puzzle. It soon became clear that Gene Griffin, along with all of his organization who survived the transition period, would belong to AT&T.

Meetings were held with craftsmen, telling them they would have the choice of staying with Southwestern Bell or going with AT&T. Shortly thereafter, meetings were held, telling these same craftsmen that they would have no choice.

Meetings were held with supervisors, telling us our wishes would be taken into consideration. If anyone wanted to stay with Southwestern Bell, every effort would be made to see that this came to pass. It soon became clear that if you asked your name was mud and you would be open to ridicule.

Business accounts were assigned to technicians so that each individual was responsible for seeing to it that his/her customers were happy. Technicians were told the man-hour performance index (that most coveted index of all) was out the window. We would be measured only by customer satisfaction and increased profits.

Our men and women would install and move telephones for their assigned customers, correct any trouble in their systems, and if a good deal of our employee's time was spent in enhancing customer relations, so much the better.

Most of our technicians thought this was one heck of a good idea. They liked it so much that the company soon decided the plan must be flawed. Scrap the plan; save the index. Let's get back to man-hour performance. We can't live without it.

As the end drew nigh, AT&T decided they had way too many employees. If they were going to be competitive, they had to be "lean and mean", an expression that we were to hear countless times.

Southwestern Bell didn't want more folks so the Company offered incentives to technicians and supervisors with seniority if they would take early retirement. Some did, but as time passed, it became obvious that AT&T wanted to be even leaner and meaner.

The date of December 31, 1983, was to be Ma's last day of life. It drew ever nearer. For many of us, it was as though we were on death row with a loved one while awaiting the arrival of the executioner.

NINETEEN

THE END

December 30, 1983. On this last work day of the year, one day before Ma Bell pulled her last breath, I retired. Thirty-two years, four months and twenty-two days after boarding a bus to Clinton and another to Sayre, I said my good-byes and, with mixed emotions, rode off into the sunset. At the age of fifty-one years, one month and two days, I was only one of many whose name cluttered the pension rolls.

By leaving when I did, I was able to retire from Southwestern Bell and not have my pension come from AT&T. Not that this was earth shattering news, but it seemed appropriate. Even though we had been a subsidiary of AT&T, we in Southwestern Bell had our own personality and our own loyalties.

Besides, had I stayed with AT&T, I would have eventually been sent to Denver or New Jersey or some such place.

My first day after retirement was a sad one. We attended Lloyd Blasingame's funeral. He with whom I had pole-buddied so many years back, and with whom we had become such good friends, had fought the good fight but even so had been overwhelmed by a debilitating illness.

Ma too died quietly. There was hardly a whimper when at midnight on December 31, 1983, she passed from this world at the venerable age of 107. Although in the months ahead, many would pray for her resurrection, unlike the Phoenix, Ma would never rise from her ashes.

Whether or not divestiture of the Bell System was justified or has been beneficial to the consumer, must be left for others to decide. Perhaps there isn't any yes/no answer. It is difficult for me to be objective. When your *Ma* dies, you naturally feel sorrow.

Like it or not, maybe the old girl's time had come. In this changing world, not all actions of governments, corporations, or influential individuals make sense to us who plod along, hoping for a better tomorrow. Neither the individual nor corporation is immortal.

When you're one of a million, there is no way you can become acquainted with all of your siblings. Still, I came into contact with a number of people and most of the memories are good.

With over twenty years in management I've lost count of all those who once reported to me. Most were good, dedicated employees.

As a craftsman, I had seven different bosses; as a supervisor, fourteen. These twenty-one individuals had much in common. They all believed in Ma and in her dedication to customer service.

At the risk of sounding maudlin, I mourn not only the demise of Ma Bell but also for the multitude of her children who have joined her in death. Old Boog and Slick and Smiley, Countryman, Steed, Bill Wells and Joe Bobo--The Telephone Man, have all departed from this earth. So have Jerry Suva, R. O. Wall and Melvin Taylor, along with numerous others. While hunting, Comer's heart quit beating. Sure he had his shortcomings, as do we all, but he was one dedicated telephone man.

Not long after retiring, I was a pallbearers at Jay Worden's funeral. Some years ago, Bill Kincaid suffered a stroke and was forced to take medical retirement. Just recently Bill died and I heard it was from a heart attack. Yesterday we were young men. Today...

Come now, enough of that. Change is inevitable and so why not accept it and plunge on.

Recently I walked over the farm where I spent my childhood. This is where I had seen the young men walking wood and where the decision had been made to put in my application for employment with Ma Bell.

There had been a time when I thought our way of life on this farm would last forever. Had the setting changed? Yes, most definitely.

During my childhood, seven people, including my grandmother and grandfather, had farmed the land, made a garden, milked cows, and worked at all the assorted functions that were then necessary for survival. There had been two houses on the farm, along with an assortment of outbuildings.

All that remain are the remnants of a rock shed and the concrete floor of a smoke-house. The old houses have been moved away. Our garden spot is grown up in trees. The windmill's head has long ago blown off. The land now belongs to a fine family who own a number of farms. They are called DAVIS FARMS. To succeed, they have had to change with the times.

As I strolled over the ground where once I'd dreamed dreams, my eyes became moist. It was here that

I had reached maturity, and here is where Dixie and I had lived when Ted was born. Here is where I had tried for one year to be a farmer on my own.

My heart ached for those innocent years, the more simple times when we all await the call that will too soon decide our destiny. We have a tendency to remember only that which is good about those days. This is as it should be.

Reluctantly I walked away and experienced the painful knowledge that the river will never flow backwards. That history of our past is forever written. Even the events of yesterday can never be changed.

Change? Yes. Ma Bell, the farm, my family and myself: we must all move ahead, changing usually in small degrees, although there are times when we seem to take colossal leaps.

It has been said that hope beats eternal. Thank God for that. We cannot change one millisecond of the past, but we can, in our unique way, help to shape the future. Would we have it any other way?

Isn't that part of what life is about? Shouldn't we just keep plugging away and do the best we can until it is time for someone else to write the chapters?

Who killed Ma Bell? This is a difficult question to answer. Although the Government spent vast sums attempting to tear the old girl down, and although it is this same Government who has been dancing on her grave, there is a chance that had the trial continued Mr. George Saunders and his staff could have come to her rescue.

Did AT&T welcome divestiture? It's beginning to look as though she did. The evidence seems to point in that direction. If so, perhaps it was a wise decision.

There is no doubt but what the parts are worth more than what the whole would have been.

Sure, there's competition. The Companies that spun off of Ma Bell, however, are in a strong position to compete and are able to enter into markets previously forbidden. When you combine net worth and profit for the Companies, all have risen dramatically since breaking apart. Although I haven't run the numbers, it is reported that there are some who are nearing a thousand percent increase in stock value appreciation.

Yeah, some of the Bell stockholders are probably doing a little dancing on the old girl's grave.

Did anyone kill Ma Bell? The fact remains that a consent decree was reached. The question is, did AT&T welcome the decree or did those in charge consent only to salvage the best deal possible?

We can only speculate.

Who killed Ma Bell? In my heart, I want to blame Uncle Sam. Who knows? Perhaps the old girl committed suicide.

EPILOGUE

We appreciate Walt Brauer who has provided us with firsthand information about the breakup of AT&T. Not only that but he has provided us with interesting insights into the Bell System. Walt was on Company Headquarters Staff for the four years immediately prior to divestiture, so he was privy to information not available to most of us.

Walt says, "Readers Digest, in the fifties, I think, had an article about the Bell System symbol, i.e., the bell within the double circle (or was it a double circle?). A man, preparing to establish a telephone answering business, was looking for a trademark. He discovered to his profound amazement that the Bell System symbol had not been registered in the U. S. Patent Office.

"They settled out of court for a paltry (I think) $17,000. I don't remember why they settled for so little; I think they were originally asking for much more."

Something that many of us had heard only as rumors, is now confirmed by Walt when he tells us, "In 1979, AT&T set out to establish the State of Texas as a separate Bell Operating Company. Southwestern Bell was the largest Bell company (in terms of assets). In 1980, several functions (pertaining to Texas) were moved from St. Louis to Dallas in preparation for establishing the new Company. But as divestiture talk (in the

Department Of Justice) became more serious, the Texas Bell plan was put on hold."

Like many of Ma's children, I have in the past blamed Judge Greene for the old girl's death. As to Judge Greene's role in the disintegration of the Bell System, Walt says, "There is widespread misinformation, even among Telephone Pioneers as to the role of Judge Greene in the divestiture of AT&T. It seems that most telephone people blame him; I say, Thank God for Judge Greene. Not for the divestiture, but for recognizing and correcting some of the inequities in the original out-of-court settlement between AT&T and the Department of Justice.

"In the original out-of-court settlement, coin telephone service and yellow page advertising were to remain with AT&T. It was Judge Greene's opinion that these items were the proper domain of local exchange, not that of a long distance carrier.

"Judge Greene was assigned by the Department of Justice to monitor the out-of-court settlement to assure that the provisions of the agreement were adhered to. In some quarters, Judge Greene's actions are seen as a more evenhanded treatment of the principles in the case."

Walt also says, "'Baby Bell' is really a misnomer, considering the immensity of these newly launched companies.

"January, 1996, was the twelfth anniversary of the divestiture of AT&T. During this time, for those interested in the financial aspects, SBC's corporate stock appreciated 896%, almost doubling the 453% increase of the S&P 500 Index. Only Ameritech at 927% exceeded the growth of SBC."

Thanks, Walt, for this information.

No wonder stockholders are now and then caught dancing on the old girl's grave.

Some dozen years or so after Ma's death, much has changed and yet there are situations that remain basically the same. For all this time and until this day, Judge Greene has kept a degree of control over the fates and fortunes of the splintered companies. However, the *Baby Bells* are beginning to appeal many of the Judge's decisions and often winning. Recent legislation has also changed the court's edicts somewhat.

Although other companies, even before divestiture, have been able to and did compete against the Bell Companies, now these same Bell offshoots are entering into other markets that were not long ago forbidden to them. Given their size and managerial expertise, they should be able to compete.

There are those in the media today who refer to AT&T as "Ma Bell." To them, the Baby Bells are Ma's children. Nothing could be further from the truth.

There are eight different corporations (ten now and more are coming; others are combining so it's difficult to keep count) who fight one another as well as all the newcomers and latecomers for market share. No love is lost between these Companies who were once siblings. In a competitive environment old loyalties often fall by the wayside.

"Ma Bell, Baby Bells." It makes great media copy and that's about it. Again, repeating what our Chairman of the Board, Charlie Brown, said in 1979, "Ma Bell is a symbol of the past. Mother Bell simply doesn't live here any more." She continued to live beyond that for some of us who wouldn't let go. But there comes a time when one has to face reality.

Doing so is not easy. As Tinkerbell recently said, "As far as I'm concerned, Ma Bell is not dead; she will never die. She may have retired, but she will never die. She retired with honor."

Yes, so long as her kids are alive and continue to congregate, she will live in our memories. We will not be caught dancing on her grave, but now and then we who are stockholders might smile about it all just a little.

Ma Bell is dead, gone from this earth forever. She died a quiet death at midnight, December 31, 1983. May she rest in peace.

Good-bye, Mother Bell; thanks for the memories.